Harald Rennhofer

In-situ-creep of carbon fibres

Harald Rennhofer

In-situ-creep of carbon fibres

In-situ investigations of the structural change of carbon fibres during high temperature creep

Südwestdeutscher Verlag für Hochschulschriften

Impressum/Imprint (nur für Deutschland/ only for Germany)
Bibliografische Information der Deutschen Nationalbibliothek: Die Deutsche Nationalbibliothek verzeichnet diese Publikation in der Deutschen Nationalbibliografie; detaillierte bibliografische Daten sind im Internet über http://dnb.d-nb.de abrufbar.
Alle in diesem Buch genannten Marken und Produktnamen unterliegen warenzeichen-, marken- oder patentrechtlichem Schutz bzw. sind Warenzeichen oder eingetragene Warenzeichen der jeweiligen Inhaber. Die Wiedergabe von Marken, Produktnamen, Gebrauchsnamen, Handelsnamen, Warenbezeichnungen u.s.w. in diesem Werk berechtigt auch ohne besondere Kennzeichnung nicht zu der Annahme, dass solche Namen im Sinne der Warenzeichen- und Markenschutzgesetzgebung als frei zu betrachten wären und daher von jedermann benutzt werden dürften.

Verlag: Südwestdeutscher Verlag für Hochschulschriften Aktiengesellschaft & Co. KG
Dudweiler Landstr. 99, 66123 Saarbrücken, Deutschland
Telefon +49 681 37 20 271-1, Telefax +49 681 37 20 271-0, Email: info@svh-verlag.de
Zugl.: Vienna, University of Vienna, Diss., 2008

Herstellung in Deutschland:
Schaltungsdienst Lange o.H.G., Berlin
Books on Demand GmbH, Norderstedt
Reha GmbH, Saarbrücken
Amazon Distribution GmbH, Leipzig
ISBN: 978-3-8381-0061-6

Imprint (only for USA, GB)
Bibliographic information published by the Deutsche Nationalbibliothek: The Deutsche Nationalbibliothek lists this publication in the Deutsche Nationalbibliografie; detailed bibliographic data are available in the Internet at http://dnb.d-nb.de.
Any brand names and product names mentioned in this book are subject to trademark, brand or patent protection and are trademarks or registered trademarks of their respective holders. The use of brand names, product names, common names, trade names, product descriptions etc. even without a particular marking in this works is in no way to be construed to mean that such names may be regarded as unrestricted in respect of trademark and brand protection legislation and could thus be used by anyone.

Publisher:
Südwestdeutscher Verlag für Hochschulschriften Aktiengesellschaft & Co. KG
Dudweiler Landstr. 99, 66123 Saarbrücken, Germany
Phone +49 681 37 20 271-1, Fax +49 681 37 20 271-0, Email: info@svh-verlag.de

Copyright © 2009 by the author and Südwestdeutscher Verlag für Hochschulschriften Aktiengesellschaft & Co. KG and licensors
All rights reserved. Saarbrücken 2009

Printed in the U.S.A.
Printed in the U.K. by (see last page)
ISBN: 978-3-8381-0061-6

Jedes Seelenbild ist auch ein Weltbild. Selbst das realistischste Weltbild ist auch ein Seelenbild.
(H. Kräftner an H. Eisenstein)

Science may set limits to knowledge, but should not set limits to imagination.
(Bertrand Russel)

Abstract

Carbon fibres are the most favourable material for industrial applications at high temperatures and high loads. They exhibit outstanding mechanical properties at low weight, which are maintained even at high temperatures due to the strong covalent bonding of the carbon atoms within the graphene sheets. Thus, the fibre properties and their structural development at high temperatures is of particular interest for technical applications.
Within the frame of this thesis, two different testing devices were developed and constructed, which allowed simultaneous loading in vacuum and *in-situ* structural investigation by X-ray scattering. One of these devices was designed for testing of fibre-bundles in the laboratory equipment (a rotating anode generator), the other for testing of single fibres in a synchrotron radiation source, which were carried out at BESSY (Berlin).
The goal of these experiments was to correlate the long-term mechanical behaviour, in particular the creep-like elongation of fibres at high loads and at high temperatures, with the structural change within the fibres, to clarify the underlying structural process and to determine numerical values for the mechanics-structure relation. This should improve the fundamental understanding of the so-called stress graphitisation in carbon fibres.

Structural investigation comprised the evaluation of the inter-layer spacing d_{002}, the crystallite size L_c and the orientation of the graphene planes from the azimuthal spread of the d_{002}-reflection from wide-angle X-ray diffraction and the size and orientation of pores within the fibres from small-angle X-ray scattering. Mechanical tests delivered the stress-strain curves. All the parameters were recorded in dependence on time for different temperatures and different loads. It was observed that the main part of the mechanical elongation and the structural change occurred in a very short period after heating up to the final temperature, which is frequently called primary creep. This is followed by a steady-state regime with a nearly constant strain rate. Our experiments enabled the determination of the activation volume and the activation energy, which is very helpful to distinguish different structural processes being responsible for stress graphitisation.
Additionally, the thermal expansion of the carbon nanocrystallites within mesophase-pitch based fibres was studied in this thesis by heating up fibres and measuring simultaneously the lattice distances *in-situ* in a synchrotron radiation source.

With this thesis, a new model for the structural process of stress graphitisation could be developed and is supported by highly sophisticated mechanical experiments.

Zusammenfassung

Kohlenstoff-Fasern zeichnen sich durch herausragende mechanische Eigenschaften aus, die auch bei Temperaturen bis über 2000 °C und unter Spannung erhalten bleiben. Das macht sie zu einem der begehrtesten Materialien in der Luft- und Raumfahrtindustrie. Daher ist es von besonderem Interesse das Hochtemperaturverhalten von Kohlenstoff-Fasern unter Last zu kennen. Im Rahmen dieser Arbeit wurden zwei verschiedene Versuchsanordnungen entworfen und gebaut, um die Struktur von Kohlenstoff-Faserbündel bzw. einzelner Kohlenstoff-Fasern mit Röntgenstreuung *in-situ* bei hoher Temperatur in Vakuum zu bestimmen. Die Versuche wurden im Labor (mit einer Röntgendrehanode) bzw. am Synchrotron BESSY (Berlin) durchgeführt.

Das Ziel dieser Arbeit war es, Informationen über das Kriechverhalten bei hohen Temperaturen und hoher Last mit Strukturveränderungen in der Faser zu verknüpfen, um den zugrundeliegenden Prozess zu finden und dafür aussagekräftige Messwerte zu bestimmen. Diese Informationen sollten das fundamentale Verständnis der sogenannten Spannungs-Graphitierung von Kohlenstoff-Fasern verbessern.

Bei den Strukturuntersuchungen wurden folgende Parameter bestimmt: Der Ebenenabstand d_{002}, die Kristallitgröße L_c und die Orientierung der Graphen-Ebenen bezüglich der Faser-Längsachse aus dem Signal der Röntgen-Weitwinkelstreuung und die Orientierung und die Größe der Poren aus der Röntgen-Kleinwinkelstreuung. Alle Parameter wurden in Abhängigkeit von der Zeit des Kriechprozesses, von verschiedenen Temperaturen und von verschiedenen Spannungen gemessen. Gleichzeitig wurde jeweils eine Spannungs-Dehnungskurve aufgezeichnet. Die Fasern zeigen Kriechverhalten, was zu einer permanenten Veränderung aller gemessenen Parameter führt. Ein großer Teil der Veränderung geschieht dabei in der kurzer Zeit nach Aufheizen der Probe, während des primären Kriechens. Danach folgt eine konstante Veränderung bei konstanter Kriech-Rate. Es konnten die Aktivierungsenergie und das Aktivierungsvolumen bestimmt werden, was es erlaubt auf verschiedene dem Kriechen zugrunde liegende Prozesse zu schließen. Zusätzlich wurde die thermische Expansion von mehreren Pech-basierten Fasern, insbesondere den Fasertypen K137 und FT500 untersucht. Dazu wurden die Gitterkonstanten der aus Graphenschichten aufgebauten Kristallite in den Fasern bei verschiedenen Temperaturen *in-situ* mit Synchroton-Strahlung gemessen.

Die vorgestellte Arbeit ermöglichte es ein neues Modell für die durch Spannung induzierte Veränderung der Faserstruktur zu entwerfen.

Contents

1	**Introduction**	**1**
2	**Theory**	**4**
	2.1 Carbon fibres	4
	2.1.1 Carbon and its allotrope - graphite	5
	2.1.2 Structure	7
	2.1.3 Production	9
	2.1.4 Properties	10
	2.2 Creep	12
	2.2.1 Dorn equation	13
	2.2.2 Characteristic parameters	14
	2.3 WAXD	14
	2.3.1 Bragg-Equation	16
	2.3.2 Interaction with matter	16
	2.3.3 Diffraction of Carbon fibres	19
	2.3.4 Data evaluation	21
	2.4 SAXS	22
	2.4.1 Guinier Radius	24
	2.4.2 Porod regime	24
	2.4.3 Scattering of carbon fibres	25
	2.4.4 Data evaluation	26
3	**Experiments with carbon fibre bundles**	**28**
	3.1 Experimental setup	29
	3.1.1 The vacuum vessel	30
	3.1.2 The sample holder	32
	3.1.3 Clamping of the bundle	33
	3.1.4 Heating of the fibre	35
	3.1.5 Temperature measurement	37
	3.1.6 Calibration of the pyrometer	40
	3.1.7 X-ray transparency	43

Contents

   3.1.8 Advantages of the setup 46
   3.1.9 Companies 47
  3.2 The experiments 47
   3.2.1 The creep threshold 47
   3.2.2 The number of fibres in the bundle 48
   3.2.3 Sample preparation 52
   3.2.4 Experimental details 53
   3.2.5 Measurement procedure 54
   3.2.6 The measurements 56
  3.3 Results 66
   3.3.1 Mechanical Parameters 66
   3.3.2 The orientation of the graphene layers 68
   3.3.3 The inter-layer spacing 72
   3.3.4 Radius of gyration 74
   3.3.5 Size of the graphene crystallite L_c 77
   3.3.6 Thermal expansion coefficient 78
   3.3.7 Chemical analysis 80

4 Experiments with single carbon fibres 82
  4.1 Experimental Setup 84
  4.2 The experiments 87
   4.2.1 Conducting adhesive 87
   4.2.2 Calibrating the stress 87
   4.2.3 Single fibre resistivity 89
   4.2.4 Calibration of the temperature measurement 89
   4.2.5 Measurement procedure 99
   4.2.6 Overview of the single fibre measurements 101
  4.3 Results 104
   4.3.1 Structural change during in-situ creep 104
   4.3.2 Thermal expansion coefficient 106

5 Post-Creep measurements 113
  5.1 Experimental procedure 113
  5.2 Results 114

6 Discussion 116
  6.1 Thermal expansion of nanocrystallites 116
  6.2 Creep of carbon fibres 119
  6.3 Structural model 123

7 Conclusion		**128**
7.1 Outlook		129
Appendix		**131**
List of Figures		**137**
List of Tables		**138**
Bibliography		**145**
Acknowledgement		**147**

Chapter 1

Introduction

Carbon fibres are the *miracle material* [24] of the old as well as the new century. These fibres exhibit outstanding mechanical properties, with a tensile strength much higher than steel, a density much less than aluminum and a Young's modulus close to diamond. The reason is the highly anisotropic structure, the strong bonding of carbon atoms within the plane and the weak bonding out of plane. The fibres are able to withstand extremely high temperatures (in vacuum more than 2000 °C) with nearly no change in the mechanical properties [71]. Carbon fibres are the basic reinforcing material in various modern technical and industrial applications, in particular for aeronautic and space industry. They are also used as the reinforcing component of composites in sports [12], e.g. frames of cars, bikes and tennis rackets, but also in architecture [21, 17].
Carbon fibres are designed to match specific applications [25] by variation of the mechanical properties (density, tensile strength, elongation at break, tensile modulus), which cannot be optimized simultaneously [81]. The temperature dependence of the Youngs modulus and the tensile strength, thermal properties (conductivity and expansion) of the carbon fibres as well as the tensile strength and the bending strength of composite materials using carbon fibres as reinforcing component have been investigated [24]. The main test methods to determine the fracture strength are the single fibre [65, 77] and the fibre bundle [51] tension test. The latter gives a statistical distribution of the fracture strength of thousands of fibres, which is often described by the unimodal Weibull distribution, but sometimes a bimodal Weibull distribution [63] is necessary.

The long term mechanical behaviour at high temperatures and at high loads is of particular interest for the application in aerospace industries. An overview on the creep of ceramics (e.g. Al_2O_3, SiC, MgO, ZrO_2) in general is given by [15, 16]. Only few papers can be found in literature, which deal

with the creep of carbon fibres, e.g. creep tests on carbon composites and carbon fibres have been performed [75]. Friction stress, creep parameters and activation energies have been evaluated for carbon composites and are compared to the values obtained for matrix free carbon yarn [76]. Knowledge of the activation energies and a corresponding activation volume allows to determine the creep process [45]. The final heat treatment during the production process, and thus, also the creep process can affect the properties (i.e. the Young's modulus) of the fibres significantly [55].
The reason for the change in the fibre properties due to creep was found in the change of the fibre structure. Carbon fibres are about $7\,\mu$m in diameter and are made of graphitic crystallites elongated along the fibre axis and crosslinked perpendicular to it. The properties of the fibres depend on the orientation of these crystallites and of the pores enclosed with respect to the fibre axis.

First post creep investigations compared tensile strength, degree of graphitisation and the crystallite size within the carbon fibres before and after the creep process [46]. The graphene layer orientation increases with increasing heat treatment temperatures [56]. Also the degree of graphitisation increases, and thus, the spacing between the graphene layers within the crystallites approximates the value of graphite due to the process of carbonization and graphitisation.
Structural investigations were frequently performed with SEM and TEM [44], finding the stacking height of the graphene planes and the interlayer spacing reduced by creep. SEM is typically used for the visualization of fracture surfaces, which give some insight into the cross-sectional texture of carbon fibres [44].

> In fibres based on polyacrylonitrile (PAN) a random arrangement of the carbon layers within the fibre cross section was usually observed, while for mesophase-pitch (MPP) based fibres the internal structure is known to depend strongly on the fibre production process. Pronounced non-random arrangements of the carbon layers within the fibre cross section, such as radial- or onion-skin structures, were reported for MPP-fibres (Edie & Stoner, 1992).[1]

Further knowledge about the fibre structure was gained by wide-angle X-ray diffraction (WAXD) and small-angle X-ray scattering (SAXS) [61]. Methods to evaluate the stacking height of the crystallites and the dimension of microvoids have been proposed [78] and results showed increasing

---
[1]This part is cited from [58].

stacking height and increasing dimension of the microvoids with increasing temperature. The voids in MPP-fibres show higher orientation along the fibre axis than the voids in PAN based fibres [33]. Additionally, the crystallite size and the degree of graphitisation are precisely determined from X-ray scattering [56]. A model for the structure of a carbon fibre, built of elongated ribbons with a wrinkled cross section texture is proposed by [55]. Crystallite growth and an increasing degree of graphitisation with increasing creep temperatures were observed in systematic studies using a rotating anode X-ray generator on PAN based fibre bundles [67], after the creep process.
The enhancement in X-ray diffraction techniques allows more detailed studies of carbon fibres using synchrotron X-ray sources. A synchrotron radiation microbeam was applied for the first time at the ESRF [58] to investigate the cross section of single carbon fibres with a position resolution in the range of one to a few microns. MPP-fibres showed an arrangement of the carbon layers different from PAN based fibres. The axial and cross sectional orientation of the carbon crystallites were measured in the same way and compared to values obtained from fibre bundle tests showing that the orientation of the crystallite with respect to the fibre axis is higher compared to results from single fibres. Furthermore a skin core structure was found for some PAN-based fibres [58].

Until today most structural investigations have been performed on untreated carbon fibres or on carbon fibres after a high temperature treatment. No information from the literature on the time dependence of the creep process is known to the author. The aim of this thesis was to follow the process of creep within the fibres time resolved. Carbon fibre bundles and single carbon fibres have been investigated during the creep process with an enhanced laboratory X-ray diffraction equipment and with a synchrotron radiation microbeam, respectively. The change of the structure of the carbon fibres was observed *in-situ* in dependence on time at different temperatures and loads applied to the fibres.

This thesis is structured as follows: Chapter 2 gives a description of the material used in the experiments and an overview of the basic theory of creep and of X-ray diffraction. In addition chapter 3 describes the experiments performed with carbon fibre bundles and chapter 4 the experiments on single carbon fibres, respectively. Finally in chapter 7, a conclusion and an outlook are given.

Chapter 2

Theory

This chapter includes the theoretical background, which is fundamental for the chapters 3 and 4, where a description of the experiments and the results are given. After an overview on the production, the structure and the properties of carbon fibres (section 2.1), section 2.2 outlines the mechanical background of creep tests and the procedure of data evaluation. Finally, wide- and small angle X-ray scattering, synchrotron radiation and the specific information, which one obtains from diffraction patterns of carbon fibres are found in the parts 2.3 and 2.4, respectively.

2.1 Carbon fibres

This section summarizes the properties of carbon fibres. It is based on the fundamental papers from Dresselhaus and coworkers [20], Fitzer [26], Walsh [81] and own previous work [66].

Carbon fibres are endless fibres consisting of more than 90% carbon atoms and have a diameter of about 7 to 10 μm. They are obtained from pyrolysis of different precursors, where the most widely used precursor is the polymer polyacrylonitrile (PAN) [27]. High tenacity (HT) fibres are usually obtained from PAN precursors. A second important precursor is mesophase pitch (MPP), which is a liquid crystalline phase of carbon and allows the production of carbon fibres with extremely high Youngs modulus. This is of particular importance for the construction of light-weight structures. Further precursor materials are cotton, bamboo or rayon, which are rather cheap, but are only rarely used due to the superior properties of MPP and PAN fibres.

Other types of carbon fibres are the only some nanometers long, but highly organized, ex-polymer filament called *whisker* or continuous filaments produced by CCVD[1]. Similarly, *carbon nanotubes* are elongated fibre-like structures from pure carbon, with a tube-like atomistic structure instead of a plane-like as the fibre types mentioned above. These types of carbon fibres are not a theme of this thesis.

Carbon fibres are often used as reinforcement materials or as fibre/resin composites with *strength-to-weight properties superior to those of any other material* [20]. The origin is the highly anisotropic bonding within the fibre structure. The fibre is built of elongated graphite like crystallites oriented along the fibre axis. The typical hexagonal structure of the graphene sheets is caused by covalent bonding of the atoms within the plane, whereas the out of plane bonds are comparably weak Van der Waals bonds.

PAN fibres are known to usually have a very high tensile strength $(3.1 - 4.6\,\text{GPa})$, a high Youngs modulus $(230 - 260\,\text{GPa})$ and low density $(\sim 1.8\,\text{g}\,\text{cm}^{-3})$ [20]. MPP fibres usually have a lower tensile strength $(2.2 - 3.5\,\text{GPa})$, but a very high Youngs modulus $(660 - 960\,\text{GPa})$ and density of $(\sim 2.2\,\text{g}\,\text{cm}^{-3})$ [20, 81]. There are also PAN fibres with a higher Youngs modulus (e.g. $\sim 380\,\text{GPa}$ for the HTA2400 fibre) and MPP fibres with a low Youngs modulus (e.g. $\sim 200\,\text{GPa}$ for the K321 fibre), respectively. The quality and the properties of the fibres can be further improved or affected by post production treatment. All fibres are usually coated with a polymer. This procedure is called *sizing*.

The development of carbon fibres already started in 1860 by fabricating ex-bamboo and ex-cotton fibres. Since the 1960's, carbon fibres were the object of extensive research mainly done in Great Britain, Japan and the USA. Further steps of development were the ex-rayon fibre, the ex-PAN fibre (cheap in production) and the pitch based fibre (ultra high Youngs modulus). Since the 1980's, the consumption of fibre materials increased drastically. Today about $27 \cdot 10^6$ kg/a of carbon fibre tow are produced worldwide [68], including MPP, PAN and ex-rayon fibres. An increase of 26% is expected until the year 2010. The pitch-based carbon fibre capacity will nearly double until then. They are used as follows (dates from 2006): 28% aerospacedefense industries, 50% industry (including infrastructure, wind power and oil and gas) and 22% sports goods.

2.1.1 Carbon and its allotrope - graphite

Nowadays, a large number of carbon allotropes is known. The carbon atoms form a zero dimensional 0D molecular nanostructure such as fullerenes, a 1D

---
[1]Catalytic chemical vapour deposition

2.1. Carbon fibres

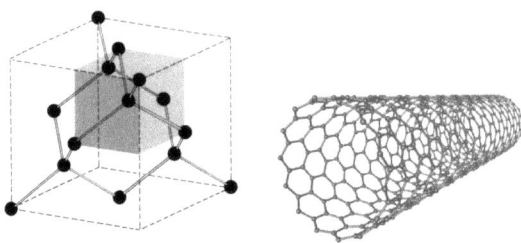

Figure 2.1: Carbon atom arrangement in the diamond structure on the left hand side [42] and a carbon nanotube on the right hand side [5].

structure as carbon nanotubes or linear acetylenic carbon (carbyne), a 2D molecular nanostructure such as graphene sheets ([28], [53]), or a 3D solid such as turbostratic or crystalline graphite or diamond. Which allotrope is formed depends on the electronic bonding. The electronic configuration of carbon atoms is $1s^2 2s^2 2p^2$. In the plane like structures (graphite, graphene) the electronic configuration is altered by the sp^2 hybridization where three free electrons are in covalent bonding (σ-bonds) and the fourth electron is delocalized in a weak Van-der-Waals bonding (π-bond).[2] This leads to two dimensional layers with atoms in a strongly bonded hexagonal arrangement interlinked with weak π-bonds. Differently, in diamond the electronic configuration (sp^3 hybridization) leads to a tetrahedral configuration in which all four electrons reside in covalent bonds. This gives diamond its outstanding hardness and leads to a high band gap, which is characteristic for an insulator.

The ideal stacking order in graphite is usually A-B-A which is also called the *Bernal structure*.[3]

The crystal structure with the unit cell is depicted in figure 2.2. The numerical values of the lattice constants are, according to [48]:
$$a = 2.459 \, \text{Å}$$
$$c = 6.70 \, \text{Å}.$$
The distance between two layers (i.e. A-B), which is usually measured by X-ray diffraction, is $3.35 \, \text{Å}$ and the in-plane bond length (side of a hexagon) is $1.41 \, \text{Å}$.

---

[2] The dislocated electron causes the good electric conductivity of graphite.
[3] Other types of order like A-A-A or A-B-C-A are called graphite like.

2.1. Carbon fibres

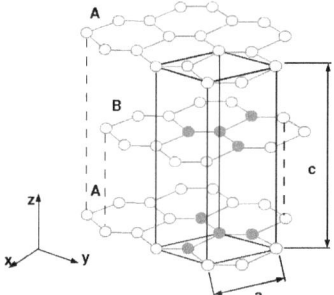

Figure 2.2: Graphite lattice in *Bernal structure* showing the characteristic lattice constants a and c. The unit cell is indicated by the cuboid.

It is generally known [20] and confirmed in simulations [10] that carbon fibres do not show the ideal graphite structure. The production process leads to rather small crystallites without a regular stacking order of the graphene sheets. The interlayer distance is not constant and one speaks of *turbostratic graphite*, see figure 2.3. In this structure a mean interlayer spacing d_{002} is found, instead of a constant repeating distance as in crystalline graphite.

Figure 2.3: Scheme of turbostratic graphite layers. There is no regular stacking order. The interlayer distance is not constant.

2.1.2 Structure

Carbon fibres contain of up to 99% carbon atoms.[4] Information on the structure is gained by different complementary methods. One can cut and polish fibres embedded in a sustaining matrix and investigate the sample with a light microscope. This is useful to obtain mean values of the fibre diameter

---

[4]A chemical analysis of the fibres used during this work is presented in section 3.3.7.

7

2.1. Carbon fibres

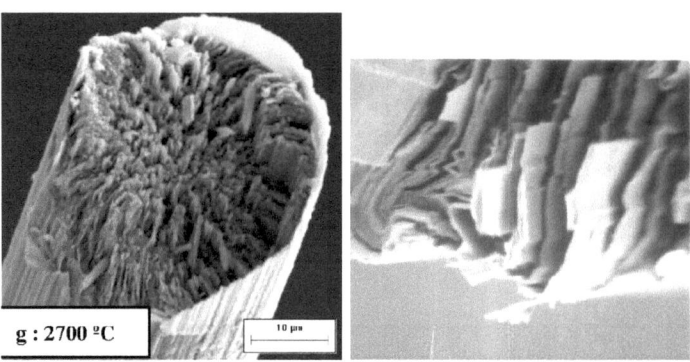

Figure 2.4: SEM pictures of fibre sections from [56]. One can see the folded crystallite sheets each consisting of some turbostratic graphene layers.

or numbers of fibres. Using polarized light also the cross sectional texture can be determined [39]. X-ray diffraction (XRD) represents an integral technique, which averages the properties in a large volume. However, with a synchrotron radiation microfocus, a position resolution of less than 100 nm is possible [58]. Transmission electron microscopy (TEM) or high resolution electron microscopy [14] give detailed information about a certain region of the fibre (figure 2.4).

Ex-polymer carbon fibres consist of elongated and rather flat ribbons of turbostratic graphite crystallites [20]. These crystallites consist of several layers of graphene, elongated along the fibre axis[5] (figure 2.5) and are only some nanometers thick [78]. The dimensions of these crystallites corresponding to the crystal structure are labeled with L_c and L_a. The high tensile strength of the fibres has its origin in the strong covalent in plane bonding of the elongated graphene sheets. Between the turbostratic graphite crystallites elongated voids, also called pores are enclosed (figure 2.6) [78]. Pores and crystallites are not oriented perfectly parallel with respect to the fibre axis, but usually enclose a certain angle with the fibre axis. This angle is usually denoted as misorientation angle [20].

Due to fabrication and heat treatment the fibres show different designs and different structures [43]. Also skin and core of the fibres sometimes feature a different arrangement of the graphene sheets [49, 50].

---

[5]The dimension of the crystallites along the fibre axis was determined e.g. for the HTA5131 PAN fibre with about 100 nm [66].

2.1. Carbon fibres

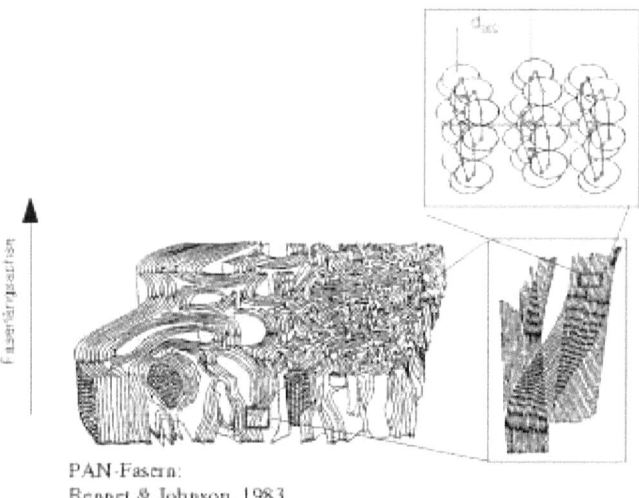

PAN-Fasern:
Bennet & Johnson, 1983

Figure 2.5: Microscopic structure of a carbon fibre [11]. Honeycomb-hexagons of carbon atoms build elongated turbostratic graphene crystallites [40].

2.1.3 Production

This section is based on information from Loidl [48] and Walsh [81]. The fibres are derived from various precursors (PAN, pitch, phenol, rayon, vinyl polymer, ...). Although the details of production may vary from fibre type to fibre type the main steps in production are similar: *spinning, stabilisation, carbonisation, graphitisation, sizing* [66]. During this process the form of the fibre is established, then the aromatic ring structures are built to make the fibre stable for the following high temperature treatment. Non carbon atoms (oxygen, hydrogen, nitrogen) are split off in inert atmosphere at more than 1000 °C. Finally the mechanical properties like stiffness, Youngs modulus and tensile strength can be increased by high temperature treatment at more than 2000 °C, which is leading to a graphite like structure, with a higher crystallite orientation along the fibre axis [20]. This is usually done for MPP fibres, but also for certain PAN fibres, and can be accomplished by tensile stress applied on the fibres. The sizing, which is usually a polymer enhancing the processability of the fibres, is highly important and can alter the fibre properties especially in carbon composites, altering the interface

9

2.1. Carbon fibres

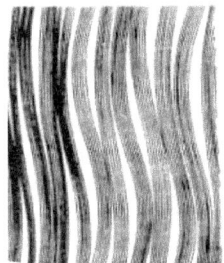

Figure 2.6: Scheme of the interlinked ribbon like structure of a carbon fibre, showing crystallites and pores [73].

between fibre and matrix or fibre and fibre [37].
Increasing the orientation of the graphite crystallites, and thus the orientation of the carbon sheets parallel to the fibre axis is one of the main possibilities for the production of high modulus fibres. In figure 2.7 this is depicted for fibres from three different precursors.

2.1.4 Properties

Physical properties

Measurements of the properties and the related values of carbon fibres can most often be compared to values obtained from experiments performed with bulk graphite or graphite foils. The values are related via indirect measurements (i.e. porosity from SAXS) [20].

- *Melting Point of graphite* [6] : $\sim 3823\,\mathrm{K}$

- *Density* [20] : $\sim 1.8 - 2.2\,\mathrm{g\,cm^{-3}}$
  The precise density depends on the fibre type and on possibly applied additional heat treatment. The higher the Youngs modulus, the higher the fibre density [20]. For high modulus fibres, the value is close to that of ideal graphite ($2.26\,\mathrm{g\,cm^{-3}}$).

- *Thermal expansion coefficient (c-direction)* [20] :
  $\alpha_{c,1200\,K} \sim 30 \cdot 10^{-6}\mathrm{K}^{-1}$
  The thermal expansion coefficient in carbon fibres is temperature dependent and exhibits large anisotropy. Bulk graphite shows the same behaviour. The values of the coefficients in a- and c- direction of the lattice are not the same. Furthermore α_a is anomaly small ($\sim 1 \cdot 10^{-6}\mathrm{K}^{-1}$)

2.1. Carbon fibres

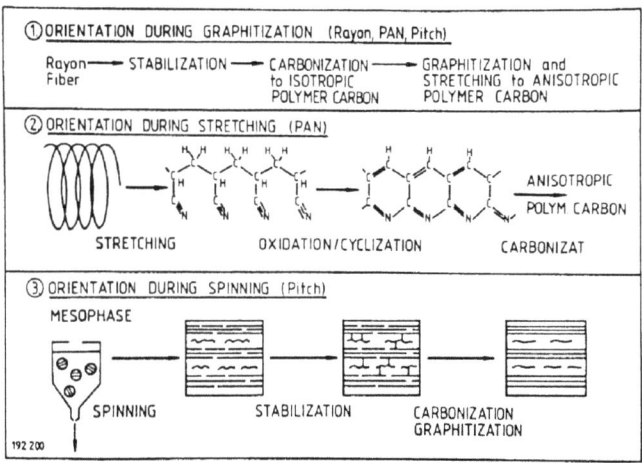

Figure 2.7: Scheme of the production process for three different precursors: rayon, PAN and pitch base, from [26]

and is even negative for temperatures below approximately 800 °C, which is well documented in literature [71, 35, 69, 79].

If α is the expansion coefficient, Δl the elongation, l the length and ΔT the difference in temperature, than the following relation holds in general:

$$\alpha = \frac{\Delta l}{\Delta T \cdot l} \tag{2.1}$$

- *Electrical Resistivity* [20] : $\sim 1 \cdot 10^{-5} \Omega\,\text{m}$
  The resistivity is temperature dependent: The fibres behave like a semi-conductor. If heat treated, the resistivity permanently decreases [71].

Mechanical properties

The mechanical properties of some widely used carbon fibres (the Youngs modulus, the tensile strength and the failure strain) are given in table 2.1. Many of the individual parameters can be controlled via the production process, but not all parameters can be optimized at the same time.

In general pitch based fibres have a higher Youngs modulus, whereas ex-PAN fibres are widely used as so-called intermediate fibres, fibres with a Youngs modulus of about 250 GPa and tensile strength of about 3 GPa. In

2.2. Creep

Fibre type	σ_T /GPa	E_Y /GPa	ϵ_{max} /%	cit
Ex-PAN: High str.	3.1-4.6	230-260	1.3-1.8	[20]
Ex-PAN: UHM	1.9	520	0.4	[20]
Ex-pitch: High str.	1.9	380	0.5	[20]
Ex-pitch: UHM	2.2-3.5	690-960	0.3	[20], [81]
HTS 5631	4.3	238	1.8	[7]
HTA 5131	3.92	235	1.7	[3]

Table 2.1: Mechanical properties of carbon fibres - typical values: σ_T... tensile strength, E_Y... Youngs modulus, ϵ_{max}... failure strain. *High str.*... high strength (also called high tenacity), *UHM*... ultra high modulus. The classification of the fibres can also be found in [81]. The fibre HTA5131 is used in this work.

table 2.1 one can see that with increasing Youngs modulus the failure strain decreases, and frequently this is also the case for the tensile strength.

Usually post production heat treatment or graphitisation under tension increases the Youngs modulus of the fibres. One can take the Youngs modulus of a single graphene crystallite in the carbon fibre (1140 ± 10) GPa as an upper limit for the Youngs modulus of the fibre itself [52]. A value obtained for bulk graphite is (1020 ± 30) GPa (from [13]), and [26] gives ~ 1060 GPa. This discrepancy of single crystal values and values obtained for bulk material are discussed in [52].

The mechanical properties of the fibres are as well temperature and stress dependent [71, 20]. Even at the same temperature, the stress-strain curve of carbon fibres is not linear and thus, the definition of the Youngs modulus should include the stress/strain interval, which was used for its measurement [64, 72].

2.2 Creep

Creep is the time- and temperature dependent permanent deformation of a material under constant load (stress), mostly observed (but not necessarily) at high temperatures. There are different mechanisms of creep [15, 16], for example gliding of grain boundaries, diffusion of atoms and growth of grains or pores. Which mechanism is relevant depends on the (activation) energy in a certain area of the material - one speaks of the *creep threshold* below which no creep occurs. The creep threshold for the carbon fibre HTA5131 is about 1300 °C at a stress of about 220 GPa. The measurement of this threshold will be described in section 3.2.1.

2.2. Creep

Heating the fibre under constant stress ($\dot{\sigma} = 0$) above the creep threshold leads to a continuously increasing elongation of the fibre due to a change in the fibre structure. A typical creep curve is shown in figure 2.8: If l is the gauge length of the fibre and Δl the elongation, then $\varepsilon = \Delta l/l$ is the fibre strain, which is followed in the creep experiment as a function of time. In region I (*primary creep*), pores originate, region II (*secondary creep*) is

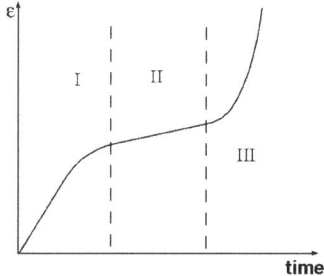

Figure 2.8: Scheme of a creep curve. Under constant load the creep strain of the fibre, $\varepsilon = \Delta l/l$ is depicted versus time. For detailed information about regions I - III follow the text.

dominated by pore growth, and region III shows appearance of cracks and pore coalescence until the fibre finally breaks. For the characterisation of the creep process, the constant creep strain rate $\dot{\varepsilon}$ in region II is used. The *Dorn-Equation* is in general used to describe this region empirically.

2.2.1 Dorn equation

If A is a constant with dimension[6], σ a constant stress, n the stress coefficient (a constant), Q the activation energy, T the temperature, and R the gas constant ($8.31441(26)$ J mol^{-1} K^{-1}), then the following relation, the so called *simplified Dorn equation* holds for the strain rate $\dot{\varepsilon}$:

$$\dot{\varepsilon} = A \cdot \sigma^n \cdot e^{-\frac{Q}{RT}} \qquad (2.2)$$

The creep behaviour in dependence on stress and temperature is described with the parameters n and Q.

---

[6]Dimension of A: $[A] = \text{N}^{-1}\text{m}^{-1}\text{s}^{-1}$.

The activation energy can be expressed in terms of the microscopic activation energy barrier H and the volume ν associated with the jump over the barrier: $Q = H - \nu \cdot \sigma$. The expression for the apparent activation volume is [34]:

$$\nu = k_B T \cdot \frac{d \ln(\dot{\varepsilon})}{d\sigma} \qquad (2.3)$$

where k_B is the Boltzmann constant $(1.380662(44) \cdot 10^{-23}\,\mathrm{J\,K^{-1}})$. The activation volume ν has to be evaluated at constant temperature.

2.2.2 Characteristic parameters

As mentioned in the last section, the creep process is characterised with two parameters:

- $n \ldots$ Tension coefficient (also creep exponent)
- $Q \ldots$ Activation energy

With the knowledge of the numerical values of these parameters, one can conclude on the underlying creep process. The activation energy Q is always related to a given volume.

For evaluating the experimental data, one takes the logarithm of equation (2.2). The relevant values of σ, T and $\dot{\varepsilon}$ are taken from the experiment, where the latter is obtained from a linear fit of the creep curve in region II. Given several measurements with the same stress σ at different temperatures, or with the same temperature T at different stresses, one can evaluate Q using the function $\ln \dot{\varepsilon}(T)$ in a linear fit or n using the function $\ln \dot{\varepsilon}(\sigma)$, respectively.

2.3 WAXD

X-ray scattering (or diffraction) is a destruction free method to reveal mainly structural information of a sample or thin film.[7]

In this section we will give a short introduction into scattering theory as discussed in [26, 42, 31]. This part also follows the short summary in [66].

After hitting a sample with X-ray photons the intensity distribution of scattered photons can be detected as a function of momentum transfer $I(q)$

---
[7]The energy transfer to the sample can be hazardous for instable, weak bonded, or biological samples, i.e. wood or protein crystals.

2.3. WAXD

(elastic scattering) or as a function of energy transfer $I(\omega)$ (inelastic scattering), respectively. In general, a triple axis spectrometer (TAS) allows the measurement of $I(q,\omega)$.

In dependence on the scattering angle, one summarizes wide angle X-ray diffraction - WAXD, and small angle X-ray scattering - SAXS. In WAXD the elastic scattering is most often used for structure analysis by Bragg-scattering, as the interpretation of information from diffuse (elastic) scattering is much more difficult, whereas the diffuse scattering is the basis of the data evaluation in SAXS. With inelastic scattering one can get information on the dynamics in a sample, i.e. molecular vibrations or the phonon density of states.

WAXD and SAXS use the same physical principle. Photons with a wavelength of about one Å (only considering hard X-rays) are diffracted from the atomic lattice planes or by inhomogeneities in the nm-range, respectively. The incoming wave-front of the photons stimulates the electrons of the sample to emit photons of the same wavelength. These outgoing photons interfere and scattered intensity is detected at positions with positive interference. It is usually assumed that all detected photons are only scattered once. X-ray photons diffracted by an atomic lattice with typical inter-lattice distances of some Å are emitted inclining the same angle as the incident photons. For structures in the nm-range the refraction angle becomes very small in the range of five degrees or less.

WAXD is described by the *Laue equations* which take conversation of momentum into account, see also figure 2.9:
Let $\vec{k}$ and $\vec{k}'$ be the wave-vectors[8] of the incident and scattered photon, then a high scattered intensity (positive interference of the photons) is observed if their difference equals a vector of the reciprocal lattice $\vec{q}$:

$$\vec{k}' - \vec{k} = \vec{q} \quad (2.4)$$

The relation of the reciprocal lattice vector $\vec{q}(hkl)$ and the correlated lattice constant in real space $d(hkl)$, with (hkl) being the *Miller indices*, is:

$$d(hkl) = \frac{2\pi}{|\vec{q}(hkl)|} \quad (2.5)$$

For infinite extended single crystals, the reciprocal lattice consists of a set of points. These reflection spots are broadened by a finite crystallite size and by atomic position defects, such as stacking faults.

---

[8]For X-rays with the wavelength λ a wave-vector $\vec{k}$ can be calculated with $\vec{n}$ being the unit vector denoting the wave-vector direction: $\vec{k} = 2\pi\vec{n}/\lambda$

2.3. WAXD

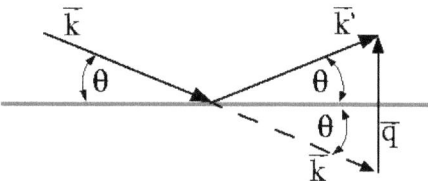

Figure 2.9: Illustration of the *Laue criterion* [66]: Momentum transfer between photon and lattice $\Delta \vec{k} = \vec{k} - \vec{k}'$ is only possible, if it equals a reciprocal lattice vector $\vec{q}$. This results in scattering of the beam under an angle of 2Θ with respect to the direct beam. Intensity is detected only under angles fulfilling this condition.

2.3.1 Bragg-Equation

When Θ is the angle between the incident beam and the lattice, n the order and λ the wavelength, then $2\,d\,sin\Theta = n \cdot \lambda$. The Bragg equation can be rewritten in terms of q:

$$q \equiv |\vec{q}\,| = \frac{4\pi}{n\lambda} \cdot sin\Theta \qquad (2.6)$$

In the transmission experiments performed in this thesis, the incident beam - and thus the non diffracted direct beam is orthogonal to the detector plane. Therefore the scattered X-rays appear under the angle $2\,\Theta$, as can also be seen in figure 2.9.

The well established convention of using $I(2\,\Theta)$ for evaluating a scattering experiment should be replaced by using the more general $I(q)$. $2\,\Theta$ is always dependent on λ and thus dependent on the setup used, whereas the use of the scattering vector q allows an easier comparison of the experimental data.

Table 2.2 gives the two most frequently used wavelengths in laboratory X-ray experiments. For comparison, in synchrotron radiation sources one often uses the same energy. However, synchrotrons allow an adjustment of the energy of X-rays in an extremely wide range (i.e. $1.9\,\mathrm{keV}$ to $30\,\mathrm{keV}$ at BESSY μSpot beamline).

2.3.2 Interaction with matter

To evaluate detected intensity patterns one has to consider the so called *transmission* T of a sample and the *background* I_{back}: If I_0 and I are the total intensity in front of and behind the sample respectively, then the Transmission $T \in [0,1]$ is defined as: $I = T \cdot I_0$, while (1-T) is absorbed. The

2.3. WAXD

source	λ / Å	E / keV
Cu-K$_\alpha$	1.541	8.1
Mo-K$_\alpha$	0.709	17.5

Table 2.2: *Source* gives the metal target used, λ is the wavelength in Å and E the corresponding energy in kilo electron volt (*keV*). For E and λ the following relationship holds, if h is Planck's constant, c the speed of light in vacuum and e the elementary charge: $E\,[\mathrm{keV}] = h \cdot c / (e \cdot \lambda\,[\mathrm{Å}])$ and thus: $E\,[\mathrm{keV}] \sim 12.4 / \lambda\,[\mathrm{Å}]$.

background is detected intensity from diffuse scattered photons and intensity not originating from photons scattered by the sample, but originating from additive scattering (cosmics, matter between sample and detector i.e. foils in the vacuum setup, etc...) or the noise of the detector. The background can be subtracted, if the transmission is known:

$$I_{\text{sample, corrected}} = I_{\text{sample, detected}} - T \cdot I_{\text{background, detected}} \qquad (2.7)$$

The transmission is determined directly or indirectly, depending on how sophisticated an experimental setup is. Measuring the intensity of the direct (not scattered) beam i.e. with a special beam-stop with and without sample one can calculate T by the fraction of both total intensities.

Most often a different method comes to account, using a material scattering much stronger in comparison with the sample (i.e. glassy carbon): If $I_{\text{sample, det}}$ is the detected intensity, $I_{\text{back, det}}$ that of the background, $I_{\text{sample + GC, det}}$ and $I_{\text{back + GC, det}}$ the intensities with glassic carbon added between beam and detector and T_{gc} the known transmission of glassy carbon, then T is given by:

$$T = \frac{I_{\text{sample + gc, det}} - T_{\text{gc}} \cdot I_{\text{sample, det}}}{I_{\text{back + gc, det}} - T_{\text{gc}} \cdot I_{\text{back, det}}}. \qquad (2.8)$$

Especially for evaluating SAXS spectra, the precise knowledge of the transmission is essential. The transmission is depending on the thickness of a sample and on the wavelength of the photons.

The transmission of carbon fibre bundles of the type used in this thesis (HTA5131) was usually about 0.90. For parts of the bundle or single fibres, the transmission is close to 1.0.

In general, the scattering intensity I transmitted after a certain length x of material with density ρ and scattering cross-section per mass unit σ with a starting intensity of I_0 is given by: $I = I_0 \cdot exp[-x\sigma\rho]$. Due to interaction with matter, σ is given by the sum of four parts: coherent scattering, photo

2.3. WAXD

effect, Compton effect and pair production. Each of these different parts of σ is depending on a certain power of the atomic number of the matter and the photon energy. The photons interact with the electrons within in the sample. Hence, the penetration depth is rapidly decreasing for heavier elements showing higher values only for nearly closed and closed shells.

In Feigin [23] the cross sections of scattering and true absorption effects are independent, represented by a sum of the two cross sections, which are each further split into a linear and a mass absorption factor. The latter being nearly constant with about $0.2\,cm^2/g$, independently from the atomic number of the material and the wavelength. Using the relation for Z the proton number and A the atomic number: $Z/A \sim 0.5$, which holds for almost all atoms.

To illustrate the complexity of the theory figure 2.10 shows the transmission for $10\mu m$ aluminium foil taken from [38] which was chosen because it is used in the *in-situ* creep experiments as a heat shield.

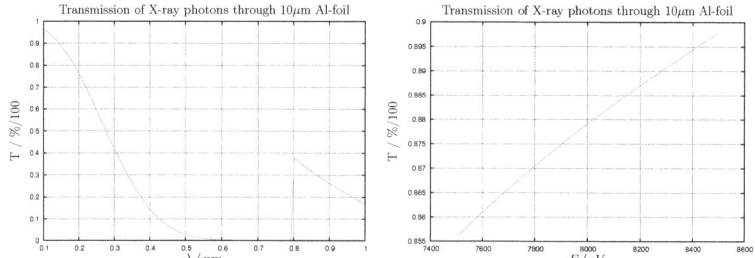

Figure 2.10: The transmission data as obtained from the web-site: http://henke.lbl.govoptical_constants. The density of Al was $2.6989\,g\,cm^{-3}$. The plots were done for $\lambda \in [0.1, 1]\,\text{nm}$ and $E \in [7.5, 8.5]\,eV$, respectively. At the wavelength $\lambda_{Cu-K\alpha} = 0.154\,\text{nm}$, which corresponds to an energy of $E = 8.1\,\text{keV}$, the transmission is about 88%. In the plot on the left hand side an absorption edge is visible. There is de facto no transmission between $0.6\,\text{nm}$ and $0.8\,\text{nm}$ wavelength.

The detected intensity is influenced by both the structure of the sample (e.g. simple cubic, face-centered cubic, hexagonal...) and the ability of the single atom to scatter. The intensity I of the diffracted X-rays by coherent scattering (diffraction) is [20]:

$$I \sim f(G) \sum_{j=1}^{n} exp(i\vec{G} \cdot \vec{r}_j) \qquad (2.9)$$

with $\vec{G}$ a reciprocal lattice vector, and further [20]:

(...) where $f(G)$ is the atomic scattering factor (the Fourier transform of the electronic charge distribution of the carbon atoms) and the sum over the n atoms in the unit cell yields the structure factor, where $\vec{r}_j$ denotes the coordinates of the atoms in the unit cell, so that the structure factor corresponds to the Fourier transform of the real space structure.

Thus, there are two factors to consider describing the total intensity of a Bragg peak. The so called form factor[9] and the structure amplitude [54].

2.3.3 Diffraction of Carbon fibres

In this section, typical diffraction patterns of carbon fibres are presented and the information obtained from these is discussed.
The structure of a carbon fibre is formed by elongated ribbons of graphene sheets, oriented nearly parallel and along the fibre axis, as described in 2.1.2. The regular inter-layer spacings of graphene planes leads to a series of X-ray reflections. In figure 2.11, one observes the 002 and the 004 reflection. From the regular arrangement of the carbon atoms within the plane, the 010 reflection is visible, which is also called 10-band which reflects the 2-dimensional origin of this reflection. The reflections are indicated with the Miller indices hkl, which also denote the related atomic planes. In figure 2.11 also the scattering vector $\vec{q}$ and the azimuthal angle χ are depicted.
A scheme of the X-ray diffraction image of carbon fibres and the corresponding structure in real space is presented in figure 2.12.

Only X-rays inclining the Bragg angle do interfere positively as given by equation (2.6). One finds two 002 patterns in a plane orthogonal to the fibre and the incident beam. The pattern is broadened in radial direction due to the turbostratic nature of the crystallites. For single crystal graphite, however, only spots would be detected, as the crystal has a fixed lattice distance with $c = 3.354$ Å [20], which corresponds in reciprocal space to a value of $q = 18.73\,\mathrm{nm}^{-1}$.
The pattern is also broadened in azimuthal direction due to the misalignment of the crystallites with respect to the fibre axis. One can think of a circle in the detector plane, thats origin is the direct beam. Then crystallites with a misalignment angle χ will show 002 intensity rotated around the beam center by an angle χ but still at the radius q given by the lattice constant.

A similar construction holds for the 10 band which lies in a plane parallel to the fibre, and orthogonal to the incident beam, originating from

---
[9]Also called atomic scattering factor or atom form factor.

2.3. WAXD

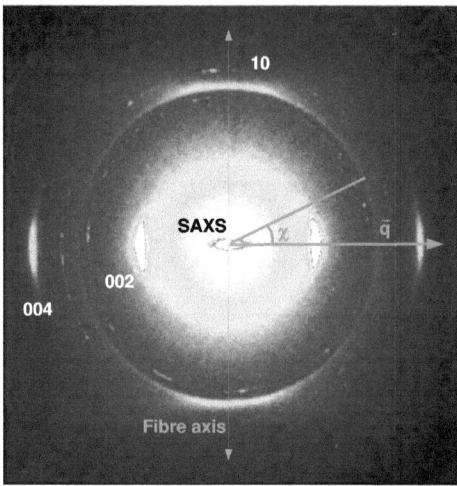

Figure 2.11: Diffraction pattern of a MPP carbon fibre. The 010, the 002, the 004 reflections and the SAXS signal around the beam stop are indicated as well as the orientation of the fibre. The scattering vector $\vec{q}$ in radial direction and the azimuthal angle χ are also depicted.

the ideal 010 reflections of the hexagonal lattice in the graphene layers. For the single crystallite the pattern consists of six spots, at symmetric positions with respect to the direct beam. Due to irregular lattice distances and misorientation of crystallites these spots are broadened in azimuthal direction but are still detected at the radius q. The huge variety of 010 orientations leads to the basically circular shape. The linewidth of this ring in radial direction is small due to the covalent in-plane bonding which leads to nearly constant inter-atomic distances in the graphene planes, and due to the large in-plane dimension of the graphene crystallites.
Scattering on the two dimensional elongated graphene ribbons result in an additional intense line, nearly tangential to the 010 reflection and orthogonal to the fibre axis, leading to higher intensity and broadening at these positions.

Onto $a < c$ the 10 band is detected at higher q values than the 002 reflections. The 004 and 006 reflections occur even at higher q values. In the experiments concluded in this thesis patterns with a higher q value than the 10 band are out of the angular range of the detector. A typical scattering pattern showing the radial or q direction and the azimuthal angle χ indicated is depicted in figure 2.11.

2.3. WAXD

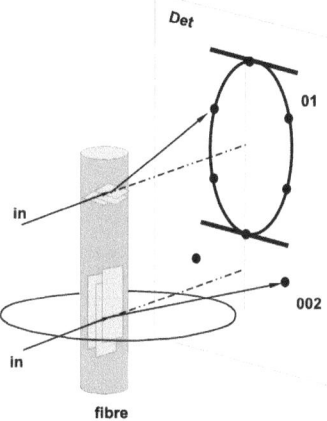

Figure 2.12: Scheme of the construction of the X-ray diffraction image of carbon fibres. The incident X-ray beam is indicated by *in*. The diffraction pattern are denoted, as well as the detector plane. The broadening of the pattern is not displayed, whereas the circular shape of the 010 reflection, originating in six spots and the 10 layer bands are visible.

One can extract the following physical information about a carbon fibre from its WAXD pattern:

- Interlayer spacing $\tilde{c} = d_{002}$ from the radial mean value of the 002 reflection.
- Mean orientation of the graphene sheets along the fibre axis $HWHM$ from the azimuthal broadening of the 002 pattern.
- Crystallite dimension L_c from the radial broadening of the 002 reflection.
- Crystallite dimension L_a from the radial broadening of the 10 band in the directions parallel and perpendicular to the fibre axis.

2.3.4 Data evaluation

Integration over the azimuth angle results in an intensity distribution depending on q; I(q), which is an asymmetric function best approximated by an asymmetric logistic function. The background was subtracted using a linear curve as fit. If I is proportional to the height of the peak, q_0 is the peak position, c is correlated to the width of the peak and d is a factor describing the asymmetry, the function is defined by:

$$Asylog(q) = I \cdot [1 + E(q)]^{-(d+1)} \cdot d^{-d} \cdot (d+1)^{d+1} \cdot E(q) \qquad (2.10)$$

Where $E(q)$ stands for

$$E(q) = exp\left[-\frac{q - q_0}{c} - ln(d)\right]. \qquad (2.11)$$

For $d = 1$ the function is symmetric, $0 < d < 1$ and $d > 1$ makes the left respectively the right part broader. The value q_0 with the maximal intensity which corresponds to $\tilde{c}$ and the full with at half maximum ω are found from the fit, the latter numerically.[10] From ω one can calculate L_c, the stacking height, or the out-of plane crystallite size (perpendicular to the fibre axis), using [59]:

$$L_c = \frac{2\pi \cdot k}{\omega}. \qquad (2.12)$$

With $k \sim 0.9$. If applying the same formula to evaluate the in-plane crystallite size $L_{a\|}$ parallel to the fibre axis or $L_{a\perp}$, perpendicular to the fibre axis, k is usually taken to be $k = 1.84$ [59].

Radial integration gives $I(\chi)$, the intensity distribution depending on the azimuthal angle χ. The data are approximated with a pdf-scaled logistic function. Let I be the intensity, $hwhm$ be the half with at half maximum of the peak and q_0 the peak position, then the function is defined by:

$$Hlogistic(q) = 4 \cdot I \cdot \frac{B^{(q-q_0)/hwhm}}{(1 + B^{(q-q_0)/hwhm})^2} \qquad (2.13)$$

with $B = 3 + 2 \cdot \sqrt{2}$ being the basis.

$hwhm$ is directly the azimuthal distribution of the crystallites with respect to the fibre axis, also called the *mean tilt* of the crystallites.

2.4 SAXS

This part is based on the following books [23, 32, 29], and on [66] - for general information on X-ray scattering see section 2.3.

---

[10]For the function can not be solved in an algebraic way for an explicit expression correlating c and d with ω.

Small angle X-ray scattering is a destruction free method to gain information about structures with different electron density, much larger than the wavelength of the diffracted photons. As a result, intensities are detected under small angles which is experimentally challenging. One can not easily shift to higher wavelengths and thus to higher scattering angles for large structures. Feigin [23] states for true absorption τ: $\tau \sim C \cdot Z \cdot \lambda^3$, where C is a constant slightly depending on λ the wavelength and Z the charge number:

> It is noteworthy that to use X rays with a wavelength greater than 2 Å in practice is rather difficult owing to the great absorption of this radiation.

The interpretation of the scattering pattern is quite different from the normal use of the Bragg equation. Although, the under-laying principles are similar, big structures are not as well ordered as atomic lattices which makes averaging over size distributions and orientations necessary. The power of SAXS is to analyse the inner structure of disordered systems, containing density inhomogeneities of colloid size (about $1 - 100\,\text{nm}$).

One can calculate the basic equations for small angle scattering. The further interpretation and solution for special particle forms and size distributions is more difficult and mostly done by approximations.

If $\varrho(\vec{r})$ is the charge density[11], $\vec{q}$ the scattering vector, ν a parameter specifying the strength of the interaction with a potential field and $A(\vec{q})$ the amplitude of elastic scattering, then the first Born approximation which is a single-scattering approximation [23]:

$$A(\vec{q}) = \frac{\nu}{4\pi} \int \varrho(\vec{r}) \, e^{i\vec{q}\cdot\vec{r}} \, d\vec{r}. \qquad (2.14)$$

In the experiment we only can measure the scattered intensity $I(q) = |A(\vec{q})|^2$. One can derive a more useful equation which is used as basic equation for further calculations. This is done assuming that the system is statistically isotropic and that no long range order exists. If D the largest dimension of a particle and $\gamma(r)$ the auto-correlation function then the intensity is given by:

$$I(q) = 4\pi \int_0^D \gamma(r) \, \frac{sin(q\,r)}{q\,r} \, r^2 dr. \qquad (2.15)$$

While $\gamma(r)$ is defined by the averaged self-convolution of the electron density distribution: $\gamma(r) = \langle \varrho(\vec{r}) \, \varrho(-\vec{r}) \rangle$. Further one can easily derive the invariant

---
[11] Also called the scattering density function.

2.4. SAXS

Q:

$$Q = \int_0^\infty q^2 \cdot I(q)\, dq \tag{2.16}$$

which is proportional to the volume V of the structure contributing to the scattering.

A very helpful way to have a first impression of the information on SAXS spectra is to have them displayed on a logarithmic scale, which allows to distinguish different shapes, such as spheres, cylinders or plates from the characteristic slope.

2.4.1 Guinier Radius

The Guinier Radius is not a real distance, but only a measure for the particle size. The formula may be found in [9], stating that averaging over possible positions around a particle leads to a correlation function given by a three-dimensional Gaussian distribution of the scattering density. In this work I follow the definition of [62], with $I(q)$ the scattering intensity and R_g being the radius of gyration:

$$I(q) \sim \frac{1}{q} \cdot e^{\frac{-R_g^2 q^2}{2}} \tag{2.17}$$

The equation is valid for $q \to 0$ and $R_g\, q < 1$. The relation of R_g to the real particle size depends on the particle shape.

2.4.2 Porod regime

Towards large q values the intensity $I(q)$ follows a power law, which is derived e.g. in [29]. Following again the notation in [62]:

$$I(q) = \frac{P}{q^4}. \tag{2.18}$$

The relation holds in the limit $q \to \infty$, if P is the Porod constant and q the scattering vector. P is proportional to the surface S of the scattering electron density. Equation (2.18) is an approximation. Although it is valid for any two-phase material with sharp smooth interfaces, it sometimes does not fit the data very well. In literature there are two ways to deal with the problem and thus to get a good fit which will give a well reproducible Porod constant. One approach to describe a non-planar surface or a density gradient is the Hausdorff-approach with a non-integer, fractal dimension, $I(q) \sim q^{-n}$, with n the fractal weight or exponent [74]. Anyhow, this was not used in this

work. The other, also used to evaluate the data from carbon fibres, can be found in [70]. Density fluctuations of the pores or the pore surface give rise to additional intensity following a square power law, and thus equation (2.18) can be rewritten as
$$I(q) = P \cdot q^{-4} + \varepsilon \cdot q^{-2}. \tag{2.19}$$
Following [60] an inhomogeneity radius l_c can be defined using the Porod constant and the invariant Q defined in equation(2.16)
$$l_c = \frac{4}{\pi} \frac{Q}{P} \tag{2.20}$$
For long elongated pores, l_c can be interpreted as value comparable to the pore diameter, and thus provides information on the pore size.

2.4.3 Scattering of carbon fibres

The SAXS signal of carbon fibres has its origin in the different electron density of the ribbon like graphene crystallites and the pores lying in between. Because of the Babinet principle one can not clearly distinguish, if the scattering signal is the one of the crystallites or of the pores. The ribbons and thus the pores are about $50 - 100\,\text{nm}^{-1}$ long and oriented along the fibre axis. The radial dimension perpendicular to the fibre axis is only of the dimension of some Å which would be some atomic layers of graphitic carbon, as was shown also in [66]. The orientation of the ribbons is not perfect, but distributed with an angle χ around the fibre axis. This effect (also visible in the WAXD signal) possibly originates partly from cross linking elements oriented in more radial direction. X-ray scattering is an integral method averaging over the whole illuminated volume and thus can not distinguish between misorientation of whole crystallites or misorientation due to highly oriented crystallites with radial crosslinking elements.

The SAXS signal is smeared out and there is also intensity found tilted by an angle χ from the ideal position. The signal is orthogonal to the fibre axis.

In the case of carbon fibres the Guinier Radius R_g is proportional to the radial dimension of the pores, perpendicular to the dimension along the fibre axis. In figure 2.13 a typical result of a scattering pattern with the intensity integrated over the azimuthal angle χ and plotted versus the scattering vector is shown. From the SAXS signal of a carbon fibre one obtains the following physical information on the fibre:

- Mean orientation of the pores/graphene ribbons with respect to the fibre axis $hwhm_p$ from the azimuthal broadening of the SAXS signal.

2.4. SAXS

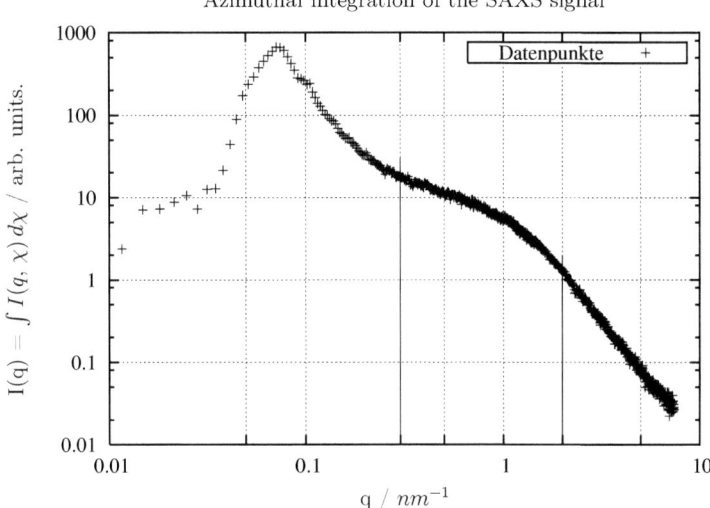

Figure 2.13: Typical plot of the SAXS signal $I(q)$ versus q. In the double logarithmic plot one can see that the intensity for high q-values follows a power law with exponent -4. Ranges with different slopes are visualized by the straight lines at $0.3\,\text{nm}^{-1}$ and $2\,\text{nm}^{-1}$, respectively. The maximum of the curve can be found at $q = 0.08\,\text{nm}^{-1}$, where the low-angle cut-off is found due to the beam stop.

- Radius of Gyration of the pores/graphene ribbons R_g.

- Values proportional to the surface, volume and radius of the pores/graphene ribbons: Porod constant $P \sim S$, invariant $Q \sim V$ and thus inhomogeneity radius l_c.

2.4.4 Data evaluation

Integration over the azimuthal angle χ leads to the intensity distribution in dependence of the scattering angle q: $I(q)$. After a subtraction of parasitic background one obtains the Guinier radius R_g from fitting the data of a plot $ln\,(q \cdot I(q))$ versus q^2 (equation (2.17)).

From $I(q)$ we also extract the values P, Q and l_c. A fit using equation (2.19) for high q-values[12] gives the Porod constant P. This is further used to calculate the invariant Q numerically [66]. The invariant Q is the area under

---
[12]In figure 2.13 for values exceeding $2\,\text{nm}^{-1}$.

2.4. SAXS

the curve $q^2 \cdot I(q)$ versus q. With a given P and Q it is easy to calculate the inhomogeneity range l_c using equation (2.20).

Radial integration gives $I(\chi)$, the intensity distribution depending on the azimuthal angle χ. The data are approximated with the function given by equation (2.13). The half width at half maximum obtained by the fit $hwhm_p$, here indicated with p for the pores, gives the mean angle enclosed between the pores/ graphene ribbons and the fibre axis.

Chapter 3

Experiments with carbon fibre bundles

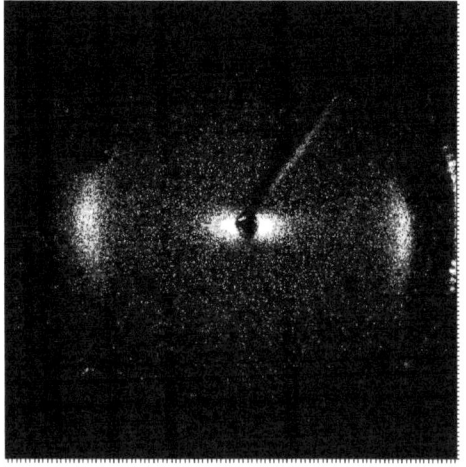

Figure 3.1: A characteristic X-ray diffraction pattern recorded of a HTA5131 fibre bundle.

There is not much literature about creep experiments on carbon fibres, and there are no publications about *in-situ* experiments, as discussed in the introduction, chapter 1. Therefore, in this thesis mechanical tests on carbon fibre bundles are combined with simultaneous X-ray diffraction measurements for the first time. This allows us to follow *in-situ* the structural change during the creep process.

In general the experimental setup has to meet the following points:

- High temperature up to 2000 °C
- Inert atmosphere or vacuum
- Tension test during two hours with constant load
- Tight gripping of the bundle
- Good electrical contact to the bundle
- Simultaneous X-ray diffraction

The apparatus constructed for the measurements with bundles complies with all required points. The setup for the single fibres does not provide information about the elongation of the fibre but complies with all other points.

3.1 Experimental setup

The setup for the experiments with fibre bundles is a further development of a setup used before in [66]. Whereas with the prototype it was only possible to investigate the fibre bundles after creep, the new setup may also be inserted into the X-ray laboratory equipment, a rotary anode generator (Rigaku equipped with a 2D position sensitive detector, HighStar from Bruker AXS). The setup consists of a vacuum vessel allowing tension tests by using a feed-through. The pressure has to be smaller than 10^{-3} mbar to prevent the carbon fibres from oxidation at temperatures above 500 °C. The bundle itself is clamped with water-cooled grips. The load applied to the fibre is measured with a load cell. The fibres are directly heated with electrical current flowing through the bundle. Thus, the grips have to be also electrically isolated from the load train and the vessel. The vessel has two X-ray windows in opposite directions made from Kapton-foil (polyimide). The window on the detector side has to be shielded with a water cooled shield due to the heat radiated from the bundle at high temperatures. The shield has a small X-ray window made of a 10 μm aluminium foil, with good contact to the cooled shield to prevent the foil from melting. The raw data for the creep curve (see section 2.2), the elongation versus time and the data from the load cell are recorded using a special program on a computer. The temperature of the bundle is measured by a two colour pyrometer of

3.1. Experimental setup

the company Keller. The signal of the pyrometer is used in a closed-loop control together with an Eurotherm process controller to guarantee constant temperature conditions during each step, where an X-ray image was taken. There are two essential points for the quality of a creep experiment: One has to know the correct temperature and one has to know the tensile stress applied to the fibre bundle. Therefore the pyrometer is calibrated before the measurement, as described in section 3.1.6. The tensile stress can be calculated if the number of fibres in the bundle and thus the exact cross section is known. The evaluation procedure to determine the number of fibres in the tested bundle is described in section 3.2.2.

3.1.1 The vacuum vessel

The vessel is mainly the supporting device for the other components of the setup. It provides the possibility of performing the experiments in vacuum and has to be vacuum-tightly attached to the laboratory X-ray device. The main requirements the vessel has to fulfill:

- Fit to the given X-ray device
- Vacuum-tight (pressure $< 10^{-3}$ mbar)
- Support for the mechanical testing device and its components
- Feed through for electricity, water and signal lines

The vessel was designed by the author and manufactured by *Pfeiffer Vacuum*. It is made of stainless steel and can be adjusted in vertical direction orthogonal to the X-ray beam. Thus it allows a rough alignment of the sample holder with respect to the X-ray beam. The vessel is planned and constructed in a way that the X-ray beam enters the center of the front flange. The sample holder is also aligned to this position. The construction of the vessel has to allow the 002 reflection of the carbon fibre bundle to be detected. For the given wavelength of copper K_α, the 002 the scattering angle is $2\theta = 27°$. For the 80 mm width of the sensitive area in the detector, this leads strong restrictions on the maximal distance of the fibre bundle to the detector. Let L be 1/2 of the width of the sensitive area of the detector and x be the distance of the fibre bundle (i.e. the scattering center) to the detector, as depicted in figure 3.3, then the following equation holds:

$$x = \frac{L}{\tan \alpha} \quad (3.1)$$

3.1. Experimental setup

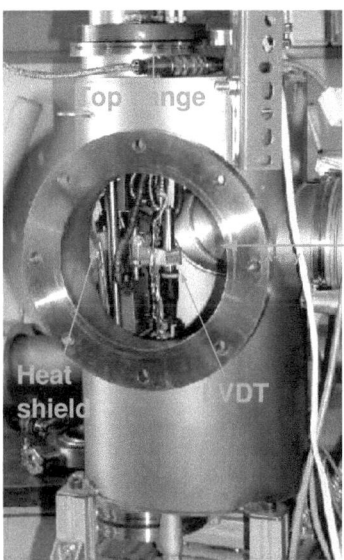

Figure 3.2: Side view of the vessel. On the right side, one can see the connection flange to the sample chamber of the X-ray device. On top of the vessel, the tension test equipment connected to the sample holder is visible. Water and electricity are fed through the upper flange. In the inner of the vessel parts, of the sample holder and the load train are visible.

Taking $L = 40\,\text{mm}$ and $\alpha = 30°$ the maximal distance is:

$$x = 69.3\,\text{mm} \sim 70\,\text{mm}$$

Thus, the fibre bundle has to be located near the front wall of the vessel, instead of the center position. On the other hand, this results in a very close distance to the front X-ray transparent window, which is fabricated from *Kapton*, resulting in the installation of a water cooled heat shield. In addition the front flange is directly attached to the vessel which requires the use of a non standard window.

The vessel can be evacuated with a rotary and a turbo molecular pump. Pressures smaller than $5 \cdot 10^{-4}\,\text{mbar}$ are usually obtained. The pressure is measured with an *Atmion compact* vacuum sensor of the company *Vacom*.

31

3.1. Experimental setup

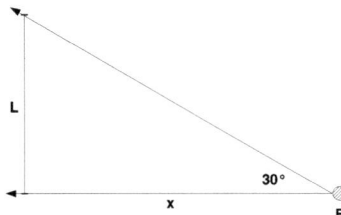

Figure 3.3: Scheme to calculate the maximum distance between detector and scattering center. B indicates the bundle, x is the distance to the detector and L is the half width of the detector window.

3.1.2 The sample holder

A scheme of the sample holder is shown in figure 3.4. A photo of the whole sample holder is shown in figure 3.5. The main flange, fixed to the vessel is used as a supporting device. The tension test equipment is fixed to the main flange and consists of hollow stainless steel rods and a cross head made from aluminium. The load cell is attached to the cross head and is shielded towards the heated fibres with a water cooled copper block. The copper block is connected to the grips clamping the carbon fibre bundle. The sample holder is completed by the second grip and the second cooling block, attached to the actuator of the tension test device using a feed through. The grips are electrical isolated using a thermal conductive silicone foil and isolating lining disks at the connection. The whole sample holder fits through a ISO-KF 100 flange.

The electrical current flows through *LEMO* connectors and isolated lines contacting the grips directly. A self designed feed through is used for the water supply. A precise description of the water circuit is found below in the section 3.1.4. The main flange, the grips, the cooling blocks and the frame are all self designed and built in the workshop of the former Institute of Materials Physics. The two 12 mm stainless steel cylinders of the sample holder are hollow tubes and are adopted to transport the cooling water. The cross-head itself is moveable, allowing the gauge length to be adjusted. If the length of the fibre is 70 mm, the center of the fibres will be exactly at the height of the X-ray beam. With a total elongation[1] of about 2 %, a cylinder stroke of about 10 mm in total is sufficient. The load cell is a 1000 N cell, signals are amplified and recorded by a computer. Cooling of the load cell is essential to prevent it from an electrical drift, induced by temperature.

---
[1] Thermal expansion and elongation due to creep strain.

3.1. Experimental setup

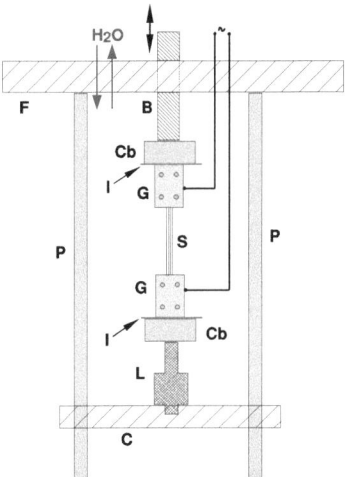

Figure 3.4: Scheme of the sample holder. F...special flange, P...steel pipes (water cooled), C...cross head, B...bulb coming from the tension testing device, Cb...cooling block (water cooled), I insulating foil, G...grips, CFB...carbon fibre bundle, L...load cell. The power lines of the current supply are indicated by the black lines. The cooling water circuit is not shown for the sake of simplicity, but the directly cooled parts appear in light blue in the scheme.

this would lead to the measurement of an incorrect force. Additionally the values do also change due to the difference between vacuum and atmospheric pressure. Thus, an offset has to be taken into account for a proper recording of the measurement data. The linear variable differential transducer (LVDT of the company *HBM*) is connected to the sample holder after mounting the fibre bundle, just before a measurement. One part is fixed to the actuator of the tension test device, the counterpart to one of the steel rods of the frame, as can be seen in picture 3.5.

The load train is perfectly aligned. The gripped fibre bundle is in one line with the center of the load cell and the center of the actuator of the tension test equipment.

3.1.3 Clamping of the bundle

The grips are made of copper to guarantee electrical and thermal conductivity. A scheme of these grips is shown in figure 3.6. A copper plate is mounted to a copper block and clamps the fibre bundle, which lies between two addi-

3.1. Experimental setup

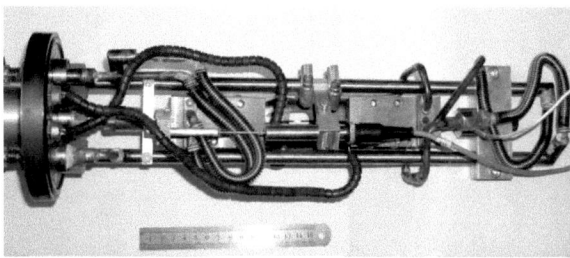

Figure 3.5: Sample holder with mounted carbon fibre bundle. One can see how the grips are attached to the cooled copper blocks linking them to the tension testing device via the plunger and the load cell respectively. Electrical current flows through the copper wires isolated by small ceramic hulls. On the upper side the displacement transducer is mounted. On the left side the feed-through and the water supply are visible. The whole equipment is fixed on the main flange plate.

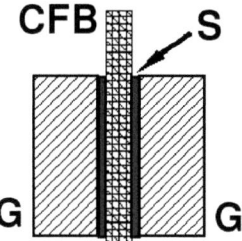

Figure 3.6: Scheme of gripping a carbon fibre bundle CFB: S...Sigraflex - described in the text, G...grip. The fibres are sandwiched between the Sigraflex to prevent them from damage due to pressure of the copper grips.

tional *Sigraflex* platelets, is clamped. The Sigraflex is a one millimeter thick carbon foil formed of several pressed layers with good thermal and electrical conductivity. It prevents the carbon fibres in the bundle from being damaged by the pressure from gripping. The copper blocks are mounted together with four screws which ensures an evenly distributed pressure on the whole cross-section. The length of restraint of the bundle is 55 mm. For this length, gripping is sufficient and no glue is required, which would lead to additional problems with conductivity of the sample at high temperatures. The limiting stress, from which on the sample starts to glide is about 550 MPa. At the ends of the grips there are inside threads allowing them to be attached to the cooled copper blocks of the sample holder. The grips themselves are

3.1. Experimental setup

not directly water cooled, because this would complicate the mounting of a sample. There are additional inside threads to attach the power supply to the grips using short imbus M6 screws. A photo of the grips together with a fibre bundle can be seen in figure 3.7.

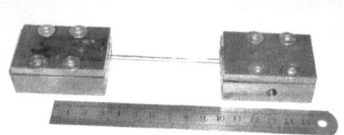

Figure 3.7: Copper grips with a mounted carbon fibre bundle. The scale indicates the dimensions. Grips with four screws have been chosen to firmly clamp the bundle. In the front of the right grip one can see a screw hole for the attachment of the power supply line, as described in the text.

3.1.4 Heating of the fibre

The fibre is a semiconductor and has therefore a relatively high electrical conductivity, which is used in the experiment for direct heating. Current is flowing through the fibres which are heated due to their resistivity. The resistivity is in the dimension of some Ohms at room temperature (RT) and decreases for higher temperatures to about 50% of the RT value at 1600 °C. The heating power can be used to control the temperature of the fibres. The temperature is measured with a two colour pyrometer PZ40 from the company *Keller*. The output signals of the pyrometer are used to control the power supply for the fibre bundle, a thyristor controlled by a *Eurotherm* process regulation unit. In addition the voltage is reduced by a transformator, which in increases on the other hand the current. The PID[2] parameters of the Eurotherm control system are as follows:

Pb = 762, Ti = 6, Td = 65

The transformator has several different contacts to be used. The voltage reduction can be chosen stepwise between a factor seven and a factor two. For the measurements usually the factors seven, five or four are used. Thus the maximal voltage of 230 V can be reduced to 33 V, 43 V or 54 V. The fibre bundle with its grips has about 3 Ω to 10 Ω and thus the current at the start of the measurement is about 11 A to 3.3 A, respectively.

---

[2] PID stands for the Proportional (Pb) the Integral (Ti) and the Derivative (Td) values.

3.1. Experimental setup

The electrical circuit is thus: thyristor - transformator - feed through in main flange - current line, isolated with ceramic hulls - grip - fibre bundle.

The heating power used in the experiments was already measured in [66], see table 3.1. A rough approximation of the part of the power radiated

T / °C	P / W
1460	40
1680	72

Table 3.1: Power needed to heat the fibre bundle. T... temperature, P... heating power. The data are taken from [66].

from the fibre bundle onto the aluminium foil gives $4.3\,\mathrm{Js}^{-1}$, which would correspond to a temperature of 660 °C. Due to the high thermal conductivity of aluminium and the good contact to the cooled steel plate, the aluminium is suited as heat shield. The heat capacity radiated from the aluminium shield to the Kapton foil is negligible.
The Kapton foil sustains temperatures up to 200 °C with sufficient strength and is furthermore cooled by air circulation.

The cooling circuit is essential for the experiment. One can see the complexity of the water cooling circuit in figure 3.5. The water is supplied by small copper pipes and flexible tubes to contact movable elements. The water circuit is thus designed by: Feed through - hollow steel rod, from main flange to lower end - cooled copper block next to the load cell - heat shield - second hollow steel rod from lower end up - cooled copper block next to the actuator - feed through.
The copper pipes and especially the cooling system of the copper blocks have an inner diameter of 3 mm. The water supply has to be clean, to prevent the system from plugging. The heat shield is built of 1 mm steel and fills the whole space between the two cooling blocks. It is cooled by a copper pipe which is soldered on the borders and has a hole in the position of the beam, where a foil transparent for X-rays is attached. This metal plate with its X-ray window was constructed to shield the whole front flange from heat radiating from the bundle. No optical path transparent for visible light or infrared radiation exists between the front flange, (i.e. the Kapton foil), and the glowing fibre bundle.
Finally it shall be mentioned, that the displacement transducer is shielded by a small stripe of aluminium foil in contact with the cooling system mounted just before the measurement.

3.1. Experimental setup

The temperature limit of direct heating

The experiments showed, that successful measurements at temperatures higher than 1800 °C are difficult and it is nearly impossible to perform experiments at temperatures higher than 1950 °C. The reason is, that fibres tend to break successively, leading to a failure of the whole bundle. Thus we were restricted to a maximum temperature of 1950 °C in our creep experiments using a carbon fibre bundle of the type HTA5131. This point was already discussed in [67]:

> However, at temperatures close to 1800 °C, the vapour pressure of carbon is rapidly increasing, which has a detrimental effect on carbon fibres not incorporated in a composite: The small volume to surface ratio of the single fibres (with a diameter of only 7 μm) and the high ablation rate in vacuum diminishes the life time significantly.

Comparison of the heating power

For the 6k fibre bundle a heating power of about 40 W for 66 mm actual length at 1460 °C was reported in table 3.1 whereas the calibration of the single fibres gives a value of about 10 mW mm^{-1} (see figure 4.7). Comparing the two values referring to 6000 single fibres per millimeter, the heating power of the bundle is about 100 times smaller than the one obtained by the single fibre. The origin of the smaller heating power required is that fibres form a bundle with a much smaller surface to volume ratio than a single fibre. Only few fibres are located on the surface, most of the fibres are in the core. A rough estimation, assuming a circular bundle, only about 270 fibres are located on the circumference and each of it exposes its half surface. Thus, only the surface of about 130 fibres out of 6000 single fibres in the whole bundle take part in the radiation to the environment.

3.1.5 Temperature measurement

The temperature is measured with a two colour pyrometer PZ40 from the company *KELLER*. It uses two wavelengths:

$\lambda_1 = 950$ nm and $\lambda_2 = 1050$ nm

measuring the fraction of the corresponding intensities. The pyrometer has a measuring spot of 3 mm at the given distance[3]. The main advantage of

---

[3]The size of the measuring spot depends on the distance of the pyrometer to the sample. In the experiment the pyrometer is located outside the window of the vessel, approximately

3.1. Experimental setup

the two colour pyrometer is that it provides the given temperature even if the sample is smaller then the measuring spot. The measurement is not affected by the glass window of the vessel or other absorbents in the optical path. The pyrometer has to be calibrated for a given material and for a certain temperature range, a will be more precisely discussed in section 3.1.6. For the carbon fibre bundles the ratio of the emissivities ε_1 and ε_2 of the corresponding wavelengths λ_1 and λ_2 was found to be

$$\frac{\varepsilon_1}{\varepsilon_2} = 101.9\,\%. \qquad (3.2)$$

Different electrical contact of several parts of the fibre bundle lead to a cross sectional temperature distribution in the fibre bundle. Usually, small fractions of the bundle, strands of several fibres, seem brighter or darker than the average and thus hotter or cooler, respectively. However, the deviations from the mean temperature in the bundle are small. Due to the characteristics of the two colour pyrometer the possibility of measuring a too low temperature can be excluded, if there is not a large fraction of cool fibres (for example if they are broken) in the optical path. The temperature deviation within the fibre bundle is estimated with about $20\,°C$ in maximum.

In addition to the cross sectional temperature distribution, a temperature gradient along the fibre bundle was found, which origins from the setup: The copper grips are heat sinks and the fibre bundle was observed to be dark (not glowing) in direct vicinity (2 mm distance) to the grips. Thus, the actual gauge length of the fibre bundle (usually 70 mm) has to be taken into account for the calibration of the creep strain rate. A temperature profile measured by a vertical temperature scan along the fibre bundle, using the PZ40 pyrometer on a z-stage, is presented in figure 3.8. The temperature decrease from the center of the bundle to its ends is found to be smaller then 3% for distances below 20 mm. For distances between 20 mm and 28 mm it is 6%, but it is even higher between 28 mm and 35 mm. One can further see the rapid decrease next to the grips.

It is likely that entanglement within the bundle is the origin for the decrease at position 10 mm, whereas the decrease at position 32 mm is due to the cooling effect of the copper block. Furthermore temperature measurements at positions between 31 mm and the grip will result in too low temperature values. This effect is caused by the dimension of the measuring spot of a diameter of 3 mm. The temperature decrease along the bundle is not changing the estimated creep rates, if the temperature is higher than the

---

20 cm from the bundle.

3.1. Experimental setup

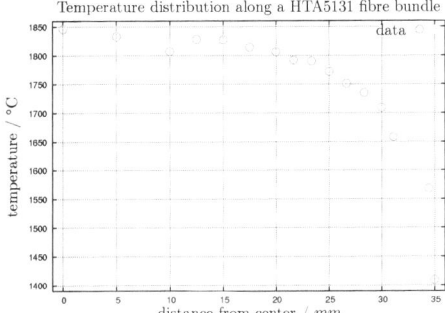

Figure 3.8: Temperature distribution along a HTA5131 carbon fibre bundle, measured with the PZ40 two colour pyrometer. The center is at position 0 mm and the end gripped with a copper clamps is at position 35 mm. The temperature decrease is smaller than 3% for distances below 20 mm, but deviates significantly from the center temperature between 25 mm and 35 mm.

creep threshold for most of the bundle along its length.[4] For the given temperature profile the mean temperature can be calculated with about 1790 °C, including temperature values measured at distances smaller than 33 mm from the center of the bundle. This is 60 °C below the maximum temperature at the center of the bundle and corresponds to a decrease of only 3.2%. The mean value increases, if only values near the center of the bundle are considered for the calculation. It would be e.g. 1805 °C considering temperatures measured at distances smaller than 30 mm.
For our experiment we conclude that the temperature decrease for smaller temperatures than 1800 °C is about 3% but at the most 5%.[5]

For the evaluation of the creep strain $\dot{\varepsilon}$ the actual gauge length is 70 mm, but taking the temperature distribution into account, the effective gauge length reduces to 64 mm.

The measurement of the temperature of the fibre bundle is performed through a glass window. Different glasses ($ZnSl_3$, $MgFl_3$, $CaFl_3$ and quartz) were considered, but quartz glass was chosen, because its transmission is nearly 100 % for light of the wavelength the pyrometer detects.

---

[4]This has only to be considered for measurements at very low temperatures: e.g. For experiments with 217 MPa at 1400 °C the temperature would be higher than 1300 °C for 62 mm gauge length - only 4 mm at each end would be cooler than 1300 °C.

[5]Assuming an higher error for smaller temperatures in realistic boundaries, the calculated activation energy at maximum increases by a factor of 1.5.

3.1. Experimental setup

3.1.6 Calibration of the pyrometer

At the beginning the limits of the pyrometer were tested by scanning across a 400 μm tungsten wire. Then the calibration of the pyrometer was performed, following a well established procedure already used in [66]: The bichromatic pyrometer uses two wavelength and can be calibrated adjusting the quotient of the two emission coefficients (equation (3.2)).

The setup

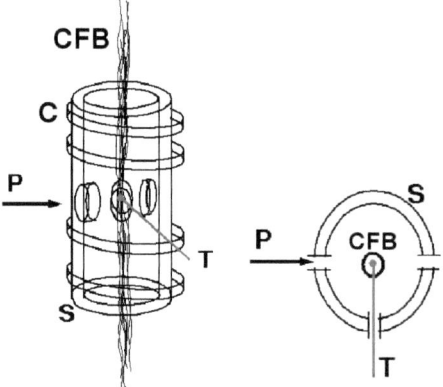

Figure 3.9: Setup for the calibration of the bichromatic pyrometer: heating of the carbon fibre bundle (CFB) and the tip of the thermocouple (T) is achieved with high frequency current flowing through the copper coil C. The susceptor (S) is induction heated and its temperature is measured with the pyrometer (P) at a distance of 30 cm.

The setup for the calibration of the pyrometer is depicted in figure 3.9. The sample is clamped at its ends, hanging through a hollow cylinder from graphite. This tube contains three drill holes on its circumference, one for the thermo couple (type B) and two aligned with the optical path of the pyrometer. The cylinder is a susceptor, which is induction heated by a copper coil, powered with high frequency current at high voltage. The susceptor is the chamber of the furnace, which heats the thermocouple and the fibre bundle. The tip of the thermo couple is next to the spot on the sample the pyrometer is pointing to. After a short period of waiting time (about 2 minutes) temperature equilibrium is reached. The cylinder wall is usually about 50 °C hotter than the sample in the tube. To avoid an error in

3.1. Experimental setup

the measurement of the temperature, two aligned holes guarantee an optical path without any influence from the tube walls. Thus, only the fibre bundle is in the optical path of the pyrometer and its temperature is measured accurately.

Additionally it is possible to position a faceplate with a hole of 3.8 mm diameter between the susceptor and the pyrometer, at a distance of 20 cm from the sample to away, to prevent from light from the outer surface of the susceptor.

The whole measurements were performed in second stage vacuum (pressure smaller then $1 \cdot 10^{-3}$ mbar), the optical path passing through a glass window. In the measurement the temperature of the pyrometer is compared and adjusted to the temperature measured by the thermo couple.[6]

Test of the pyrometer

The accuracy of the temperature measurements was confirmed by radial scanning of a 400 μm tungsten wire. This experiment was done with a constant quotient of the emission coefficient of 107.8 %. This coefficient was determined at 1100 °C, and it has to be adjusted in dependence on temperature. For example, it decreases to 104.0% for a temperature of 1400 °C. The measuring spot in this case was determined with a diameter of 2 mm. The pyrometer displayed a constant temperature for positions about 0.8 mm around the center of the wire. This leads to the conclusion, that the pyrometer gives accurate temperature values even for samples with significantly smaller size than the measuring spot.

The calibration

The measurements for the fibre bundle were twice repeated and compared with measurements of a C-C compound rod. The latter has a drill hole for the tip of the thermocouple. The data of one measurement obtained with different quotients of the emissivity are presented in figure 3.10. The temperature values measured with the quotient of the emission coefficients of 101.9 % are listed in table 3.2. The quotient for the pyrometer is obviously temperature-dependent and higher values for the quotient are obtained at lower actual temperatures. For a fixed quotient the correlation between the temperature measured with the thermo couple and the one measured with

---

[6]The thermo couple gives voltage values mV, which are transformed to temperature values using [1].

3.1. Experimental setup

U_{TC} / mV	T_{TC} / °C	T_{PY} / °C
6.25	1148	1183
7.22	1242	1268
8.48	1358	1368
9.52	1450	1450

Table 3.2: Calibration of the pyrometer: The temperatures given were measured with a thermo couple (TC) type B and the pyrometer (PY) with a fixed quotient of 101.9 % respectively. The error in the values is ±5 °C.

the pyrometer is linear, coinciding only at one certain temperature. For lower temperatures the pyrometer overestimates the temperature, for higher temperatures it underestimates the temperature.

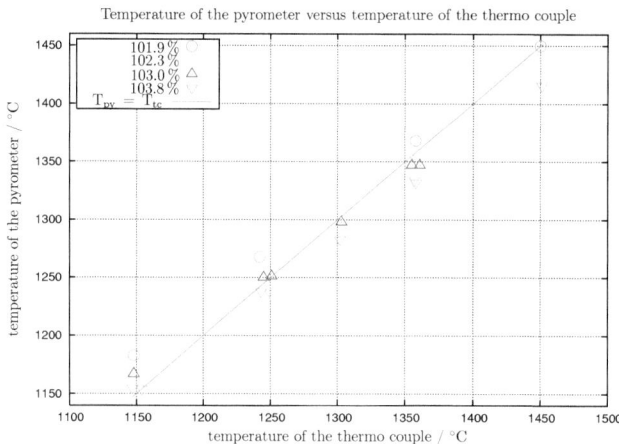

Figure 3.10: Calibration of the bichromatic pyrometer: The data for the different quotients of the emission coefficients are given. The straight line indicates an exact correspondence between the temperature measured with the pyrometer (T_{py}) and the temperature from the thermo couple (T_{tc}). The quotient decreases slightly towards higher temperatures. Several points have been measured twice: once during increasing and once during decreasing the temperature stepwise.

A quotient of 101.9 %, measured at 1450 °C was chosen and the linear correlation had to be extrapolated to 2000 °C, because our measurement setup did not allow to reach higher temperatures. A linear fit of the data with $f(x) = a \cdot x + b$ leads to: $a = (0.874 \pm 0.008)$ and $b = (181 \pm 11)$ °C.

3.1. Experimental setup

Thus, the temperature T_{tc} of the bundle is calculated from the temperature measured with the pyrometer T_{py}:

$$T_{tc} = \frac{T_{py} - 181}{0.874} \, °C \qquad (3.3)$$

A conservative estimation of the error in T_{tc} is 20 °C, considering the errors of the fit and a maximum error in T_{py} of 10 °C. The calibration is used by application of equation (3.3) during the experiment.

The quotient of the emission coefficients found for the C-C rod at 1430 °C with 100.9 % is close to the one of the fibre bundle. However, an influence of the local structure at the C-C composite could be possible. It is a 2D-composite, one plane shows the fabric with a larger amount of matrix, the other only the fibres. Therefore, the latter was chosen for the measurement of the quotient of the emission coefficient.

Measurements performed with a spectral pyrometer[7] showed, that the emission coefficient of a carbon fibre bundle (HTA5131) is approximately 80 % at 1450 °C. The coefficient is decreasing with increasing temperature.

The linearity of the function $T_{tc}(T_{py})$ for a fixed quotient of the two emission coefficients (e.g. 101.9 %) as found during the measurements is not documented in literature. As far as the theory was considered by the author, the function is not totally linear. The slope of the achieved curve is only approximately linear for quotients around 100 %. A calculation of the expected quotient as a function of temperature, based on the Stefan-Boltzmann law, gives values which are in good correspondence with the experiment: the quotients decrease for higher temperature values.

3.1.7 X-ray transparency

The path of the X-ray beam through the vessel is as follows: Kapton foil window - carbon fibre bundle - aluminium foil of the heat shield - Kapton foil - beam stop. The path between the second Kapton foil and the detector is in air, which leads to a small diffuse background. To suppress this background the beam-stop is positioned very close to the Kapton foil. The chosen setup guarantees a high scattering intensity of the carbon fibre bundle together with a small background intensity. Furthermore, the Kapton foil near the

---
[7] The intensity of only one wavelength is used and the emission coefficient itself has to be adjusted during calibration.

3.1. Experimental setup

detector has to be shielded from the heat of the bundle.
The main background origins from scattering on air and on the Kapton foil. Aluminium foil was chosen for the heat shield, because it gives a lower background than a zirconium foil in the chosen setup, as shown in figure 3.11. Though zirconium has better thermal properties (the melting point is much

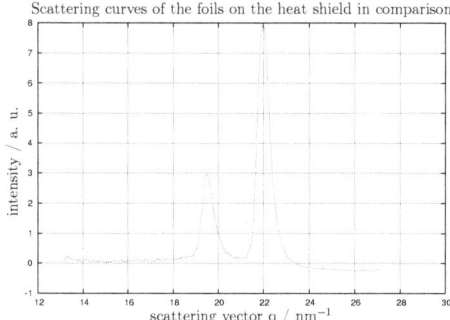

Figure 3.11: Scattering curves of the heat shields $I(q)$: The red line shows the scattering pattern obtained with a $10\,\mu$m zirconium foil attached to the heat shield. The green broken line shows the scattering pattern of a $10\,\mu$m aluminium foil with a carbon fibre bundle, intensity multiplied by a factor of four. Although the lattice parameters of zirconium are smaller than the interlayer spacing in graphite, the first zirconium peak appears at the same scattering angle as the 002 peak, because the distance specimen - detector is larger than the distance foil - detector.

higher than the one of aluminium), a $10\,\mu$m aluminium foil is advantageous, because of its low atomic weight and thus its low electron number. The relevant data are compared in table 3.3. The high heat conductivity allows

Material	M / °C	T / %	X-ray pattern
zirconium	1850	45	intensive rings
aluminium	660	88	smeared out spots

Table 3.3: Properties of the $10\,\mu$m heat shield foils: M...melting point, T...transmission. See the text for a detailed discussion.

the use of an aluminium foil because its ends are water-cooled. Even at test temperatures of 1800 °C of the carbon fibre bundle, the melting temperature of aluminium is by far not reached. The main disadvantage of the zirconium foil, however, is that one Bragg peak of the zirconium foil is detected directly

3.1. Experimental setup

at the same position of the 002 peak of the carbon fibre bundle.[8] Its integrated intensity is considerably higher than the intensity of the 002 reflex of the carbon fibre. Therefore, the aluminium foil is chosen as heat shield in the experiment.

The foil is attached to the steel heat shield with a small aluminium frame gripped with screws, which can be seen in figure 3.12.[9] The hole in the heat shield covered by the foil is elliptic (23 mm x18 mm), with the short axis parallel to the axis of the fibre bundle at a distance of 11 mm. The minimum size of the hole in the steel shield is calculated from equation (3.1).

The Kapton window on the detector side is constructed as follows: It has to be very flat and thus the 0.1 mm thick Kapton foil is glued to a steel ring fitting to the ISO-K 100 flange with an inner diameter of 80 mm and 2 mm thickness. Usual two component epoxy glue from *UHU* was used.[10] This ring can be fixed with four clamps to the flange. The tensile strength of Kapton is 230 MPa at room temperature, and 139 MPa at 200 °C, its maximum elongation at break is 70 %.

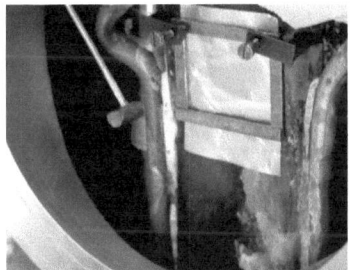

Figure 3.12: Beam-stop: The 1.5 mm thick lead cylinder is glued to a 2 mm thick copper wire and can be adjusted via a xy-stage. In the background the heat shield with the 10 μm Al-foil is visible.

---

[8]The lattice distances are quite different, but the foil is closer to the detector and thus the peaks are at the same position.

[9]While the zirconium foil has random orientation of its crystallites, resulting in a Debye-Scherrer ring of the scattering pattern, the aluminium foil is oriented due to the production process. There are several peaks, smeared out in azimuthal direction. It was thus found to be useful to mount the foil with the long dimension horizontally at the heat shield. The scattering streaks are then found perpendicular to the 002 reflection and can be shielded with metal plates attached to the detector.

[10]The glue was brought onto the flange and was then heated to result in a thin and even liquid film. Then the ring was put face down onto the flat stretched Kapton foil.

3.1. Experimental setup

The beam-stop is mounted outside of the vessel. It consists of a small lead cylinder with a diameter of about 1.5 mm which is glued to a copper wire, see figure 3.12. Top and bottom of the lead cylinder are formed cup like to suppress background from scattering at the surface of the beam stop or at the edges of the cylinder. The beam stop can be adjusted via a xy-stage.

3.1.8 Advantages of the setup

The setup was designed for the specific experiment of investigating carbon fibre bundles at high loads and high temperatures with simultaneous X-ray diffraction. Successful experiments could be performed with the equipment. A minor draw-back is that the 010 reflection is not detected. The constructing of the load train is adopted perfectly to the requirements of the setup. The fibre bundle is mounted into the sample holder outside of the vessel and thus the whole load train is removable and small enough to pass the ISO-K100 flange.
The main advantages of the setup are the possibility of accurate mounting of a sample and the high precision of the elongation measurement. The tension testing machine using a spindle provides a very accurate control of the constant load. The threshold of the machine to react after deviation of the load value from the set point is smaller than 0.5%. This allows to determine mechanical parameters such as the strain rate $\dot{\varepsilon}$ with a high accuracy.
The X-ray transparency, i.e. the ratio sample signal to background, is high. Measurements with 300 s already give evaluable statistics.

Usually no whole fibre bundle is used, but the bundle is split into two, three or four parts. This was especially found to be useful for higher temperatures, because the electrical power and therefore the heating power is reduced. Carbon fibres in general have a high transmission for X-rays of the wavelength used in the experiment, see section 2.3.2, thus a minimal number of fibres has to be used in the tests. It turned out to be useful to have a number of about 1500 to 2000 fibres, which meets the requirements of low heating power and sufficiently high scattering intensity.
A main drawback in X-ray scattering of carbon fibre bundles in comparison to single fibres is the angle distribution of the fibres in the bundle with respect to the load axis. This gives rise to a higher orientation-distribution (correlated with the angle χ) as described in 2.3.3 and 2.4.3 for WAXD and SAXS respectively. This was already described in the literature [57]. On the other hand the high number of fibres allows to use rather thick Kapton windows and an X-ray beam with a diameter between 0.5 mm and 1 mm, which leads to an increase of scattering volume and thus scattering intensity.

The angle distribution of the fibres within the bundle and additional misalignment of small strands of the fibre bundle do not alter the mechanical parameters of a tension test performed with the fibres, because the number of misaligned fibres is negligible compared to the large number of fibres parallel to the loading axis.

3.1.9 Companies

Company	Material
Pfeiffer vacuum	vacuum vessel
Trinos vacuum	vacuum components including a glass flange
Vacom	pressure gauge
Keller-HCW	pyrometer
Lemo GesmbH	connectors and feed throughs
Rattay	corrugated hoses
Metal supermarkets	steel and copper pipes
RS components	electronics and base frame of the vessel
HTM	displacement transducers
Rose+Krieger	xy-plate
Zrunek and Kurt Koller Kugellager	o-rings
Alfa Aesar	zirconium foil

Table 3.4: The table gives a list of companies, from which material and components for the experimental setup and the realisation of the measurements were acquired.

3.2 The experiments

3.2.1 The creep threshold

The creep threshold as described in section 2.2 is the point, where the elongation starts to increase significantly in comparison to the thermal expansion. In this work the creep threshold is defined in a more applicable way: *The minimum temperature for which creep occurs at a certain constant stress is understood as creep threshold.*

Two measurements have been performed to determine the creep threshold using a hydraulic tensile testing machine as described in [66]. A stress of (217 ± 1) MPa was applied to a 6k HTA5131 carbon fibre bundle. The tests

3.2. The experiments

were performed in vacuum, with a pressure smaller than $1 \cdot 10^{-3}$ mbar and the fibres were heated by current directly flowing through the bundle. The bundle was stressed first and then the temperature was increased in small steps from 1000 °C to 1500 °C and from 800 °C to 1400 °C, respectively. The temperature was measured at each step using a spectral pyrometer. The elongation of the bundle was recorded by a LVDT. The recorded data of one of the experiments are plotted in figure 3.13 elongation versus temperature. A rather smooth increase in elongation is observed due to thermal

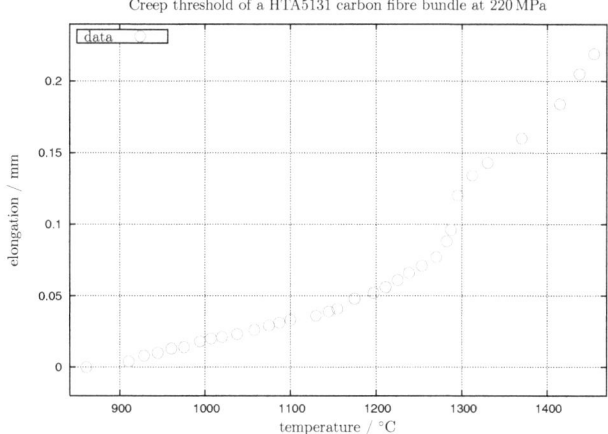

Figure 3.13: Determination of the creep threshold: The elongation was recorded for increasing temperature steps. At about 1290°C the fibre starts to creep and thus the elongation increases significantly.

expansion, followed by a steep increase at a certain temperature, which thus defines the creep threshold. From figure 3.13 the following numerical value is obtained: (1290 ± 15) °C. The value obtained from the second measurement is (1340 ± 25) °C. A mean value is calculated:

$$T_{cr} = (1310 \pm 25) \,°C$$

The creep threshold itself depends on the applied stress. For higher stress values, the fibres will creep at lower temperatures and vice versa.

3.2.2 The number of fibres in the bundle

The determination of the precise number of fibres in the carbon fibre bundle in the experiment is crucial for the calculation of the correct stress in the fibre bundle. There are three main approaches to determine the number of fibres in the bundle: Measuring a stress strain curve for the fibres, measuring the electrical resistivity of the bundle and determining the weight of a bundle of a certain size and the respective parts used for the test.

Number of fibres via the stress strain curve

With the knowledge of the Youngs modulus it is possible to calculate the fibre number from the slope of a stress strain curve [48]: If N is the number of fibres in the bundle, l the gauge length, c_{ges} the measured compliance for the system with the sample (units are m/N), c_s the compliance of the system, r the diameter of a single carbon fibre in the bundle, and E the Youngs modulus, the following equation holds:

$$N = \frac{l}{(c_{ges} - c_s) \cdot r^2 \cdot \pi \cdot E} \quad (3.4)$$

The Youngs modulus and the diameter of the of the fibre has to be known and [3] gives an overview on different measurement results.

Crucial for applying this method is the knowledge of the compliance of the load train of the experimental setup c_s. The precise determination of this compliance may be performed either with a solid block of a hard material, which exhibits only a negligible deformation, or with a stress strain curve from a well-defined specimen with known cross-section and Youngs modulus. Both approaches were chosen in this thesis using a steel bar and a carbon fibre bundle HTA5131 (6k), respectively.

Several measurements have been performed with both methods and a mean value for c_s resulted in:

$(0.91 \pm 0.07) \cdot 10^{-6}$ m/N.

Although the measurements where performed with utmost care, a control experiment with an as-received fibre bundle (6000 fibres according to manufacturer) showed that the method usually underestimates the number of fibres by a factor of two. Additionally, increasing or decreasing the stress leads to quite different results with a deviation between those two loading procedures in the range between 5 % and 10 %. *Therefore, the method is not applicable in the given setup.* One reason could be the construction of the heat shield, which is connected to the lower copper block and the frame of the sample holder with a combination of corrugated tubes and copper pipes. It is very probable, that this acts as an additional spring at low loads. The effect is overcome at high loads, which can only be applied to the steel bar

3.2. The experiments

and a whole fibre bundle near its yield stress. The test with a 6k bundle at 220 MPa resulted in a fibre number of 3700 ± 200 which deviates from a realistic number of at least 5600 fibres (a conservative estimation as some fibres break due to the preparation) in tension by 34 %.

Number of fibres via the electrical resistivity

A method with more accurate results is to measure the electrical resistivity of the bundle in the test jig. From a four point measurement from the whole fibre bundle and taking into account the number of fibres given by the manufacturer, the specific resistance of one single fibre was calculated:

$$(360 \pm 10)\,\Omega/\mathrm{mm}$$

Thus, a measurement of the resistivity of a fibre bundle and the gauge length allows the calculation of the fibre number.[11] However, a measurement of single fibres gave a resistivity of typically:

$$(420 \pm 5)\,\Omega/\mathrm{mm}$$

This shows the drawback of this method: The resistivity of the bundle is not always constant due to contact problems of several strands within the bundle. Both methods underestimate the number of fibres within the bundle, whereas the value obtained from single fibre measurements is more reliable, as it is a direct measurement of the resistivity. The error from the true fibre number, evaluated with the value gained from fibre bundle measurements, is estimated to be about 20 %. The error using the single fibre value is found to be much smaller and is about 10 %. This shows, that the resistivity method is applicable and leads to reasonable fibre numbers, but a higher precision would be favourable.

Number of fibres via the weight of a bundle

The most successful way to determine the number of fibres in the fibre bundle is to weight the bundle in advance, split it and weight the parts again. The theoretical accuracy can be calculated with the knowledge of the resolution of the scale and the weight of a test fibre bundle. E.g. the weight of a 6k bundle (13.0 ± 0.1) cm long was (88.4 ± 0.1) mg. Thus, a resolution of about 7 fibres can be achieved with a precision balance. The method was checked by weighting the split parts of the bundle and adding the individual

---

[11] As an example, for 70 mm gauge length and 14 Ω resistance, the number of fibres would be 1800 ± 50.

3.2. The experiments

results again. A representative measurement gave as sum of 3 parts of a 6k bundle 5979 fibres, compared with 4722 fibres evaluated with the single fibre resistivity method. The error of the weighting method is thus smaller than 1 %. Due to fibre loss by clamping and mounting the fibres, the error in the experiment will be about 5 %. This upper limit for the error can be estimated quite well because breaking of more than 5 % of the fibres during preparation is clearly visible and the specimen would be changed.

For fast application at the experiment the method was even further improved: the weighting method has been used to get a correlation between fibre number and electrical resistivity. Two fibre bundles have been split into four and six parts respectively and the fibre number was determined by its weight. Fibre numbers between 370 and 2200 where measured. The individual sums over the single results give (6043 ± 20) fibres for the bundle split into six parts and (5926 ± 15) fibres for the one split into four parts, respectively. Each part of a bundle was clamped afterwards, identical to the procedure of a creep experiment (70 mm gauge length) and the resistivity of the bundle was measured using a *FLUKE* voltmeter by contacting the grips. Values between $15.9\,\Omega$ and $98.5\,\Omega$ were obtained and are presented in figure 3.14. If N is the number of fibres in the bundle and R the resistivity in Ω, the following relation is determined by a linear fit:

$$N = \frac{(36640 \pm 200)\,\Omega}{R + (0.4 \pm 0.2)\,\Omega} \qquad (3.5)$$

The maximum error for N is about 4 %, which corresponds to $40-80$ fibres. The numerical value of the slope can be converted to the resistivity of a single fibre of one millimeter length, which allows a direct comparison of all methods:

$(520 \pm 5)\,\Omega\,/\,\text{mm}$ for one fibre.

In table 3.5 the comparison of all three test methods is summarized.

The weight and the resistivity method are chosen to estimate the fibre number in the experiments and to calculate the resulting stresses. Whereas weighting gives a maximum number of fibres, the resistivity method using the fit gives a good lower boundary to work with. In the example the smaller value deviates by 7 % from the higher one. The mean value is expected to be very close to the true fibre number. Thus, an error for the stress can be estimated, which will be smaller than 3 %.

3.2. The experiments

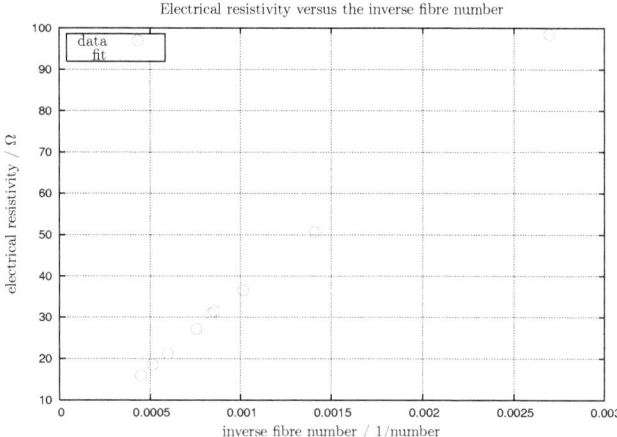

Figure 3.14: Estimation of the fibre number: The plot gives the resistivity versus the fibre number to calibrate the resistivity method. The expected linear correlation is very well reproduced in the experiment.

3.2.3 Sample preparation

The samples used in the experiments were all carbon fibre bundles or parts of carbon fibre bundles of the type

    6k Tenax HTA5131

which is produced as endless fibre bundle containing 6000 single carbon fibres (according to the manufacturer). The as-received fibres in the bundles are delivered with sizing and thus feature a very good usability and are easy to

method	measured value	fibre number
tension test increasing	-	1140 ± 50
tension test decreasing	-	1420 ± 45
resistivity bundle	$14.2\,\Omega$	1770 ± 40
resistivity s.f.	$14.2\,\Omega$	2085 ± 50
resistivity fit	$14.2\,\Omega$	2430 ± 150
weight	$38.8\,mg$	2630 ± 10

Table 3.5: Estimation of the fibre number, obtained from the different methods indicated. $s.f.$ means single fibre and fit means the correlation given in equation (3.5).

handle.[12] The specimen are cut off from a coil, on which the endless fibres are spun. The length of these specimen is at least 170 mm because the gauge length is 70 mm and each copper grip is 50 mm long.
The bundle is weighted and split afterwards, weighting each part again. The specimen is pre-loaded with 20 MPa and aligned with the grips. The bundle is carefully clamped by tightening the four screws of the grips. The resistivity of the bundle is measured by contacting the grips, then the grips are mounted into the sample holder outside of the vessel. The bundle is always handled with care, to avoid bending or break of fibres. During mounting, the fibre bundle is at no time subjected to any stress.
The current lines insulated with ceramics are then connected to the grips. The LVDT and its aluminium shielding are fixed to the sample holder, which is mounted into the vessel as a final step.

3.2.4 Experimental details

The mechanical strength of carbon fibres is statistically distributed as for many ceramics, which is in general described with the Weibull distribution. Only a large number of single fibre tests or the large number of fibres in a fibre bundle allow the determination of the strength and the strength distribution with sufficient accuracy. Likewise, the diameter of single fibres in a bundle show some variance and requires the measurement of a large number of fibres.

The carbon fibre bundle has a low heat capacity. Therefore, the temperature regulation is very sensible to variations in the net power required to heat the fibres. Small fluctuations could change the fibre temperature and thus the thermal expansion (the elongation measured). Therefore, the temperature is kept constant with a closed loop control. This guarantees that reliable values for the slope of the creep curve (the strain rate $\dot{\varepsilon}$) are obtained.

During the tensile tests usually the weakest fibres fracture even at very low loads. Their number is small for low ε, but can reach up to hundred fibres for high strain values. This would increase the strain of the residual fibres by up to 10 %, and thus change the experimental properties resulting in an increase of the slope of the stress - strain curve. Because this effect is not observed during most of the experiments a loss of a high number of fibres can be excluded.

For very accurate measurements one has to exclude thermal expansion of

---
[12] The fibre bundles were provided by *Schunk Kohlenstofftechnik*.

3.2. The experiments

the vessel or the LVDT. The vessel is at room temperature during the whole experiment, because it is constantly cooled by the environment in a climate controlled laboratory. For the displacement transducer, the critical part, exposed to heat radiation from the fibre bundle is about 5 cm long. Taking into account all relevant data (time exposed to the heat source, energy absorption, thermal conductivity, heat capacity) the thermal expansion would be 60 μm at maximum. Due to the setup of the sample holder and the positioning of the LVDT, this would lead to an offset in the same dimension as the elongation of the bundle. To suppress this effect, the LVDT is shielded with an aluminium foil attached to the cooling system for sufficient cooling. The temperature increase on the parts of the LVDT is not higher than 10 °C, which leads to an maximum offset of 9 μm.

Clamping of the fibre bundles has to be performed very cautiously to get straight and well aligned samples.
To avoid large and systematic errors in the resistivity measurements they are always done in the same way at two polished spots of the copper grips. The value is measured twice with good contact.

3.2.5 Measurement procedure

There are two different test procedures. The one is loading the fibre first to the final stress and applying current afterward. The other is to heat the fibre under a little pre-load to its testing temperature and applying the full stress thereafter. Both techniques will lead to the same results for the strain rate $\dot{\varepsilon}$, but alter the fibre response in the first five to ten minutes of the experiment. The reason for choosing the procedure of first applying the stress to the fibre is the following: When stress is applied to a heated fibre bundle, the resistivity and therefore the temperature is changed. This method would theoretically allow to distinguish between the thermal expansion of the fibre bundle and beginning creep strain, but as the experiment has shown it is not easy to resolve the beginning creep strain for fibres under high tension during heating. Thus, the specimen are usually first loaded and temperature is applied afterwards.

For each measurement several actions are required:

- Estimation of the fibre number
- X-ray scan for the bundle position
- Closing the vessel

3.2. The experiments

- Initial extension of the fibre bundle
- Start of the record of data elongation versus time
- Application of load
- Room temperature X-ray measurement
- Temperature increase to test temperature
- Successive X-ray measurements
- End of the measurement

Estimation of the fibre number: The fibre number is evaluated using techniques described in section 3.2.2. With the known fibre number the load in *Newton* for the nominal stress is calculated.[13] Therefore the fibre diameter was taken from [51] with $(6.91 \pm 0.02)\,\mu$m.

X-ray scan for the bundle position: These measurements are performed in air. A rough positioning of the fibre is done optically using the position of the beam-stop. The fibre bundle is usually loaded with 20 MPa using the tension testing machine to align the fibres. The fibre position with respect to the X-ray beam is changed stepwise by moving the main flange of the sample holder with respect to the vessel. The fibre bundle is centered by maximizing the count rate at the detector. At the bundle position usually $1800 - 2200$ counts / 5 s at 40 kV and 10 mA tube current are detected. Then, the bundle is unloaded to determine the point of zero load.

Closing the vessel: The aluminium foil is fixed on the heat shield and the Kapton foil is attached to its flange. The vessel is closed and evacuated using a rotary and a turbo-molecular pump. During the experiment the pressure is lower than $1 \cdot 10^{-3}$ mbar. The cooling water is turned on. After the load cell delivers a constant value the experiment is continued.

Initial extension of the fibre bundle: The fibre bundle is strained until the stress strain curve is linear. After unloading, the fibre bundle the values of the displacement transducer and the load cell are set to zero. Now the fibres are loaded until the nominal stress for the experiment is reached to test if fibres rupture or not. Then the bundle is unloaded again.

---

[13]This is necessary, because the controller program used for the tensile testing machine uses Newtons.

3.2. The experiments

Start of the record of data elongation versus time: The record of the stress strain curve is started.

Application of load: The fibre is loaded using the tensile testing machine until the calculated nominal value for the load is reached. The computer program controlling the tension testing machine is set to auto-control mode, which keeps the load constant by a fuzzy control.

Room temperature X-ray measurement: For comparison to the data at high temperatures, a first X-ray diffraction image of the sample is taken at room temperature. The X-ray source usually operates at 40 kV and 50 mA. To ensure linearity of the X-ray detector, the current at the rotating anode generator is increased to the maximum value which results in less than 3000 counts / s (total intensity). The X-ray diffraction image is recorded for 600 s and the same parameters are used for the following *in-situ* measurements.

Temperature increase to test temperature: Current is applied to the sample until the nominal temperature value is reached. Due to primary creep the elongation of the fibre is high in the first minutes of the experiment. The resistivity of the fibre bundle also changes fast and thus a manual regulation of the temperature was found to be optimal for the first 10 min. Then the temperature is kept constant with a closed loop control using the pyrometer signal as set point.

Successive X-ray measurements: After having reached the final temperature, X-ray diffraction images are recorded. Usually, 600 s measuring time were used, which gives at least $1 \cdot 10^6$ photons at the detector. A higher time resolution, i.e. shorter measuring times, leads to a signal, which would not allow data evaluation with sufficient statistical accuracy.

End of the measurement: After $1\frac{1}{2}$ h or 2 h, the creep experiment was stopped and the temperature was decreased to room temperature. A final X-ray diffraction image was taken *post creep* at room temperature. The sample was then unloaded, cooling water was turned off and the vessel was vented with air again.

3.2.6 The measurements

In total 16 *in-situ* creep experiments have been successfully performed at different loads and different temperatures. Most measurements show characteristic secondary creep and good statistics in the X-ray data. Each measurement was performed for a nominal temperature and a nominal stress value, which are found in table 3.6.

Measurement 4 and 6 are named twice, because the temperature (and for

T / °C	σ / MPa			
	100	170	220	260
1300	M6	M5		
1400	M1, M6	M14	M7	
1500	M2, M11	M4, M8	M9, M13	M15
1600	M3	M10	M4	
1700		M12		

Table 3.6: The table gives an overview on the carbon fibre bundle experiments, listing the numbers of the measurements. T... nominal temperature, σ... nominal stress. During the experiments *M4* and *M6* two different temperature and stress levels have been measured. M16 is not listed for it was performed at 1500 °C and 20 MPa.

measurement 4 also the stress) was changed during the experiment. Detailed parameters of the measurements are found in table 3.7.

For the measurements 1, 2 and 3 one 6k bundle was split, but the fibre numbers were only evaluated by the resistivity method (without equation (3.5)) and thus the fibre numbers (so the stresses) known during the experiment are not as accurate as for the other experiments. The numbers were evaluated after the experiments and the true stress during the experiment was recalculated. For the measurements 4, 5 and 6 an other 6k fibre bundle was split and fibre numbers were also evaluated by the weighing method, but still without equation (3.5). Again the fibre number was calculated afterwards to determine the true stress. For measurement 7 and 14 one part of a separate bundle was used and for measurements 8, 9 and 10 again a 6k bundle was split, as was done for the measurements 11, 12 and 13 and for the measurements 15 and 16, respectively. For M7-M16 the fibre number was determined accurately before the measurement using equation (3.5).

The duration of the experiments was not always the same and depended on how fast the fibre bundle was ruptured due to the creep process.
The strain versus time curves of the measurements are shown in figures 3.15 to 3.21. The values for the strain are calculated with a gauge length of 64 mm. Each curve consists of three parts: the application of the load (red part) the heating to the nominal temperature (green part) and the strain curve

3.2. The experiments

M / #	T / °C	σ / MPa	t / min	N / #	creep
1	1400	105	40	2300	×
2	1520	114	90	1200	✓
3	1600	100	80	2540	×
4	1550	200	75	2500	✓
4'	1530	190	50	2500	×
5	1300	150	110	2000	✓
6	1300	110	20	1130	~
6'	1400	110	60	1130	~
7	1400	220	40	1150	×
8	1520	160	160	1800	✓
9	1510	210	100	2400	✓
10	1600	160	30	1300	~
11	1500	140	98	1670	✓
12	1700	160	89	1220	✓
13	1500	220	104	1900	✓
14	1400	160	94	2150	✓
15	1500	260	125	1980	✓
16	1500	20	106	1240	✓

Table 3.7: Parameters of the creep experiments: M...index number of the measurement, T...temperature with an error of $\pm 20°C$, σ...stress with an error of about $\pm 3\%$, t...duration of the experiment, N...rounded number of fibres with an error of $\pm 40 - 80$ fibres, creep gives an subjective impression if the secondary creep process was visible: clear...✓, uncertain...~, not...×.

during the X-ray experiments (blue part), also indicated in the figure legends.

Figure 3.15, M1 shows a characteristic strain versus time curve where no creep was observed. The temperature of 1400 °C was well below the creep threshold for the 40 MPa applied to the fibre bundle. The *shortening* of the carbon fibre bundle due to the fact that the stress applied to the bundle is well below the creep threshold, is clearly visible.

In Figure 3.16, M4 temperature and stress had to be reduced in this test after a certain time of creep, because the bundle started to fracture. At lower temperature and lower stress, the fibre is nearly creep resistant, and only small elongation was observed.

In M6 temperature steps were performed to compare structural development due to high temperature treatment and creep respectively. It is clearly visible, that the fibre shows creep for temperatures higher than the creep

3.2. The experiments

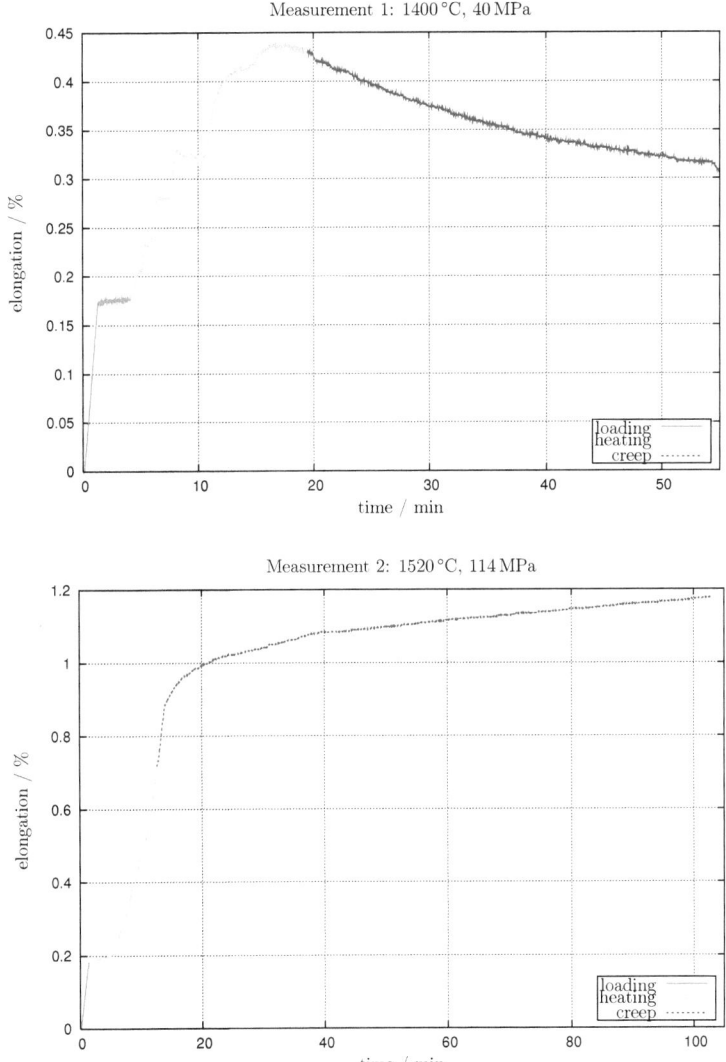

Figure 3.15: Measurements 1 and 2: At the stress of 40 MPa no creep is observed at 1400 °C. The fibres show the characteristic *contraction*. At higher stresses and temperatures creep is observed.

3.2. The experiments

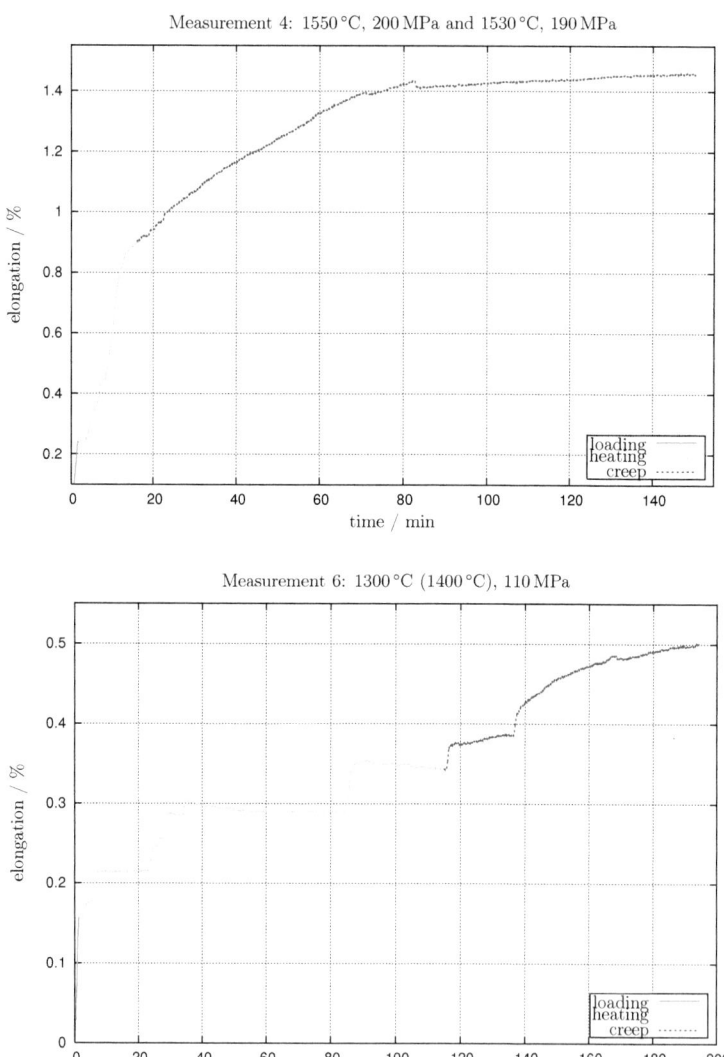

Figure 3.16: Measurements 4 and 6: At measurement 4 first the strain and then the temperature were reduced at 70 min and about 80 min, respectively. In measurement 6 the following temperature steps were applied: < 1000 °C, 1030 °C, 1200 °C, 1300 °C, 1400 °C - the two latter steps are drawn in blue colour.

3.2. The experiments

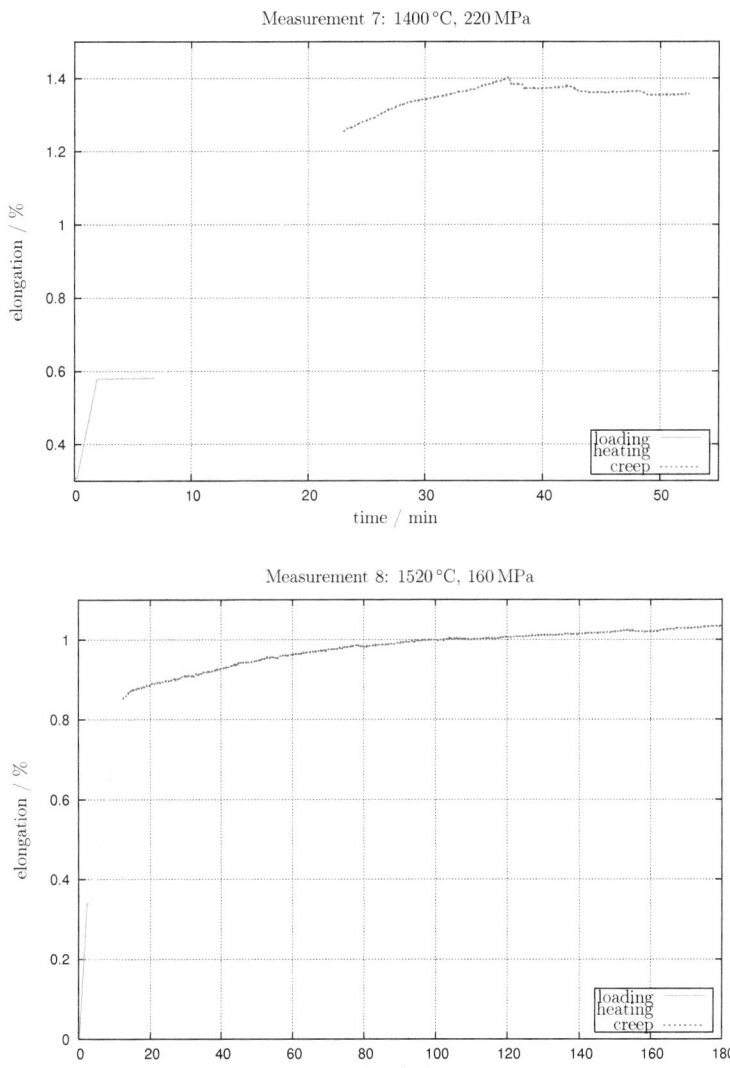

Figure 3.17: Measurements 7 and 8: In measurement 7, the curve is not smooth due to difficulties with the heating power. Measurement 8 shows a typical creep curve.

3.2. The experiments

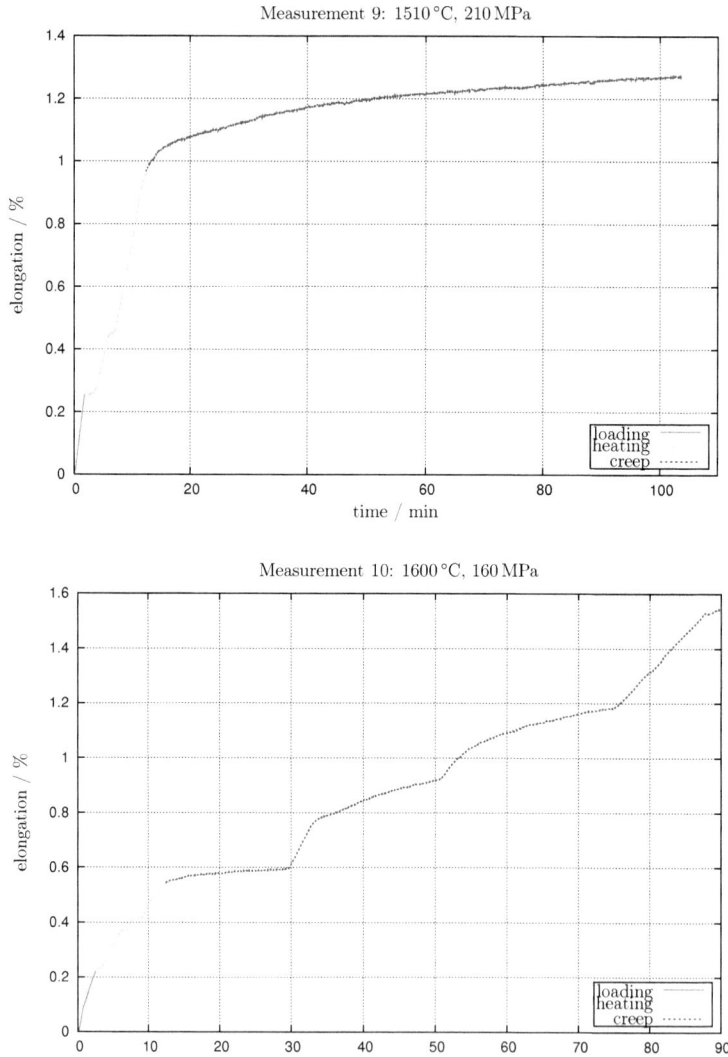

Figure 3.18: Measurements 9 and 10: Measurement 9 shows a typical creep curve. In measurement 10 the following temperature steps were applied: 1300 °C, 1500 °C, 1600 °C, 1640 °C. During the last heating step the fibres broke successively.

3.2. The experiments

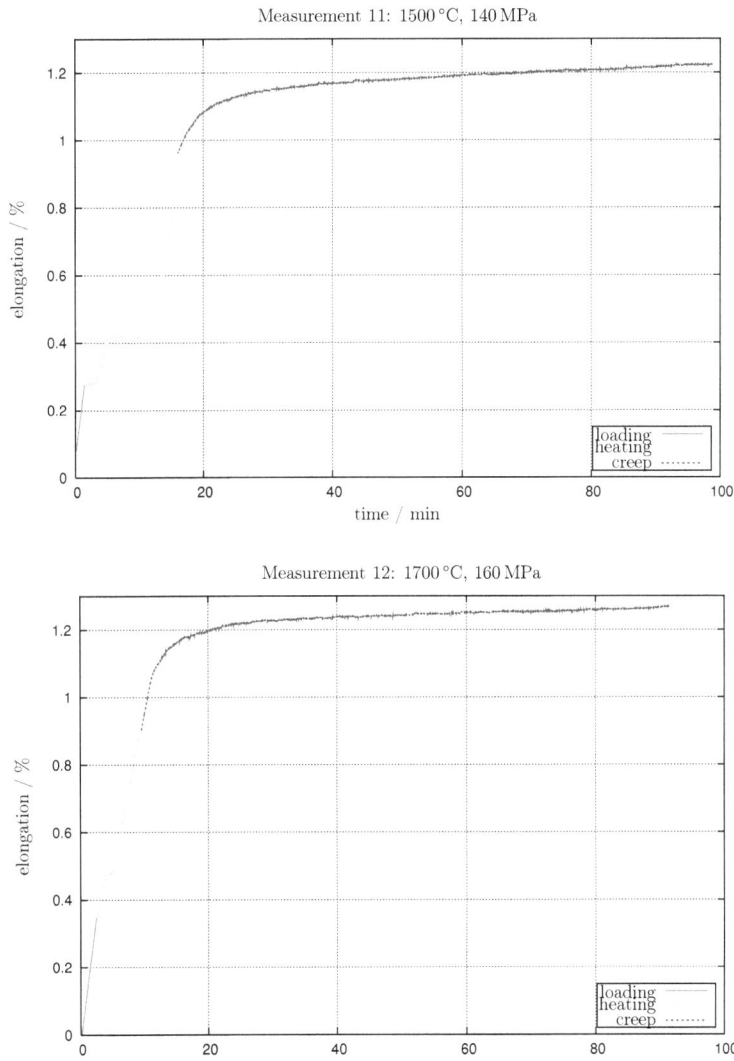

Figure 3.19: Measurements 11 and 12: M11 and M12 show typical creep curves, although the slope of M12 at steady state was found to be rather small.

3.2. The experiments

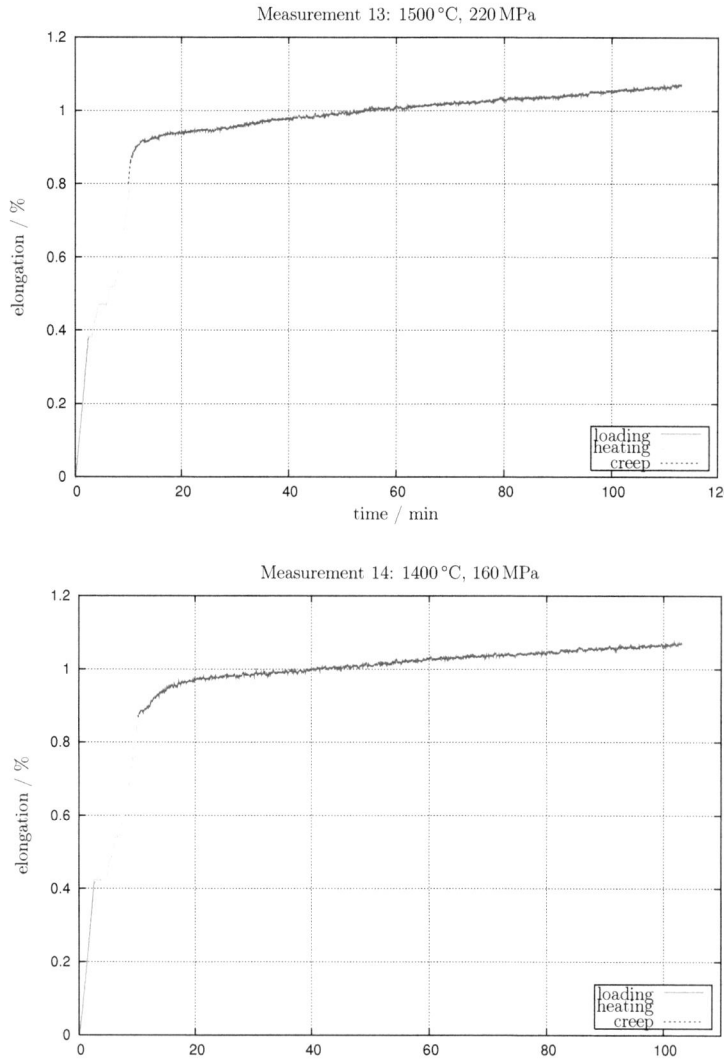

Figure 3.20: Measurements 13 and 14: Both curves show typical secondary creep.

3.2. The experiments

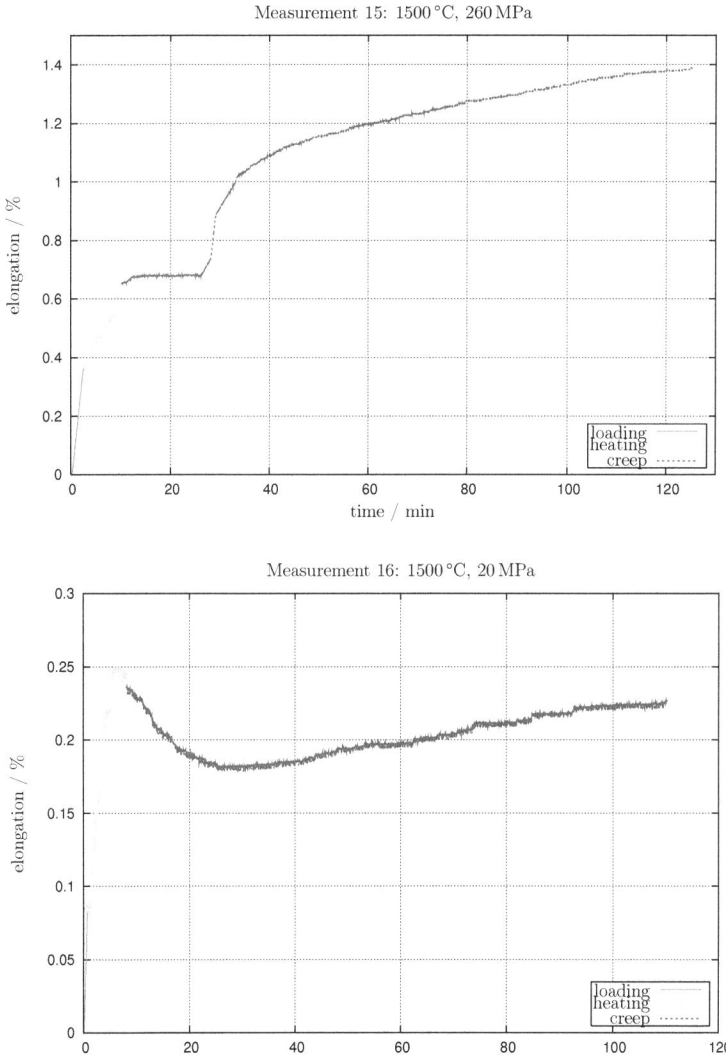

Figure 3.21: Measurements 15 and 16: The load of M15 was first chosen to be 60 MPa at test temperature and was then increased to 260 MPa. M16 shows a typical time - strain curve if the load is below the creep threshold.

threshold.

Figure 3.17: M8 shows a characteristic creep curve. Its first two diffraction pattern were recorded for only 300 s, to obtain a higher time resolution.

Figure 3.18: In M9 the first two diffraction patterns were also recorded with only 300 s for higher time resolution. This is also true for all diffraction measurements in M10. Furthermore, the temperature was increased in steps and two diffraction images were taken at each step. When the bundle started to fracture at 1640 °C, the measurement was stopped and a diffraction image was taken at room temperature.

Figure 3.19: The temperature of M12 was decreased after the shown steady state creep stepwise. After 80 min creep the temperature was reduced to 1580 °C, 1470 °C, 1300 °C and 1180 °C before room temperature. This was done to calculate the thermal expansion of the carbon fibres.

Figure 3.20: For the test M13 a small amount of fibres broke at the beginning of the experiment. This did not further influence the experiment.

Figure 3.21: To distinguish between thermal expansion and primary creep the loading procedure was changed at M15. Two X-ray measurements were performed at room temperature, at 60 MPa and 260 MPa, respectively. Then the fibre was heated with only 60 MPa stress applied and stress was increased after an X-ray pattern was recorded. After increasing the stress at M15 to 260 MPa the fibre shows usual creep behaviour. The creep curve shows constant elongation for the low stress value in time period of 15 min. This was expected because the stress is below the creep threshold. Nevertheless, it differs from the behaviour depicted in M16, where the elongation is decreasing immediately, but finally increases again.

3.3 Results

3.3.1 Mechanical Parameters

To determine the mechanical parameters of a creep experiment, it is not necessary to perform *in-situ* experiments. Although reliable results have already been presented in [66], the evaluation is performed again to compare the values obtained for the creep exponent n and the activation energy Q. Furthermore the activation volume is evaluated. Table 3.8 and table 3.9 give the values for the strain rate, determined by a linear fit in the steady state region of the creep curves. It is obvious, that the values of $\dot{\varepsilon}$ scatter considerably, which frequently the case for ceramics and was already observed before in [66]. The activation volume was determined by applying equa-

3.3. Results

Nr.	$\dot{\varepsilon} \cdot 10^{-7}$ / s^{-1}	T / °C	σ / MPa
5	2.819(2)	1300	150
14	1.628(4)	1400	160
8	7.415(9)	1500	160
10	9.26(3)	1600	160
12	0.890(3)	1700	160

Table 3.8: Evaluation of the creep curves. The number of the experiment, the strain rate in the steady state region, the test temperature and the stress during the experiment are given. The values scatter and especially the value of M12 seems to be very low. The error given in brackets is the one obtained by the fit.

tion (2.3) to data from measurements at 1600 °C, taken from [66]. The data are shown in figure 3.22 including the linear fit curve. The activation volume

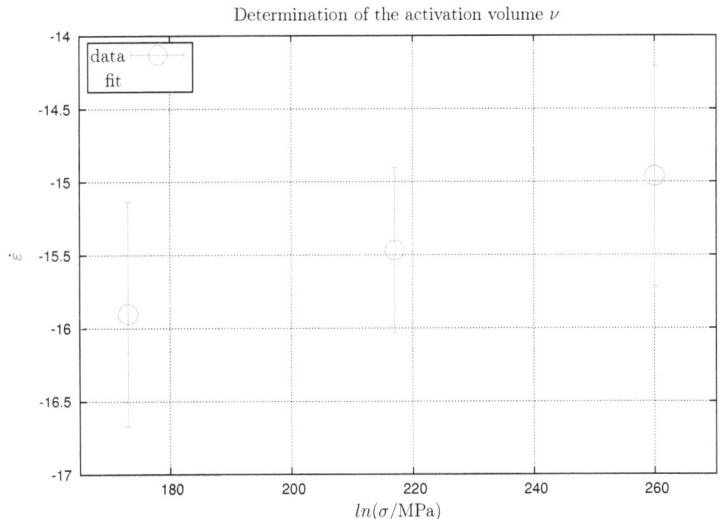

Figure 3.22: Strain rates versus stress at 1600 °C to determine ν.

was determined to be:

$$\nu = (2.8 \pm 0.1) \cdot 10^{-28}\,\text{m}^3$$

This value corresponds to a cube with an edge length of about 6.5 Å.

67

3.3. Results

The small number of the in-situ measurements is not sufficient for a statistically accurate evaluation of the activation energy Q. Thus, the activation volume obtained from the large amount of creep data from [66] was used to determine the microscopic activation energy H. This evaluation method was applied to all strain rates measured during the in-situ experiments. The mean value for H is:

$$H = (2.0 \pm 0.3)\,\text{eV}$$

This corresponds to a value of $(194 \pm 25)\,\text{kJ}\,\text{mol}^{-1}$.

Nr.	$\dot{\varepsilon} \cdot 10^{-7}/\text{s}^{-1}$	T / °C	σ / MPa
16	1.020(2)	1500	20
2	2.420(2)	1520	90
11	1.449(2)	1500	140
8	7.415(9)	1520	160
9	2.316(3)	1510	210
13	1.879(2)	1500	220
15	5.138(4)	1500	260

Table 3.9: Evaluation of the creep curves. The number of the experiment, the strain rate in the steady state region, the test temperature and the stress during the experiment are given. The error given in brackets is the one obtained by a fit.

For the creep exponent n no reliable value could be determined due to the high scattering of the data and more measurements would be required. A linear regression was performed and a value of

$$n = 0.5 \pm 0.3$$

was found. The large error shows the uncertainty due to the small number of measurements and the high scattering of data. This is also visible in figure 3.23, where the data of table 3.9 show no clear correlation.

3.3.2 The orientation of the graphene layers

The graphene layer orientation with respect to the fibre axis is depicted in figures 3.24 and 3.25. The temporal evolution of the half width at half maximum (HWHM) of the azimuthal intensity distribution of the 002 reflection during the creep process is plotted in dependence on the temperature (constant stress) and on the stress (constant temperature), respectively.

3.3. Results

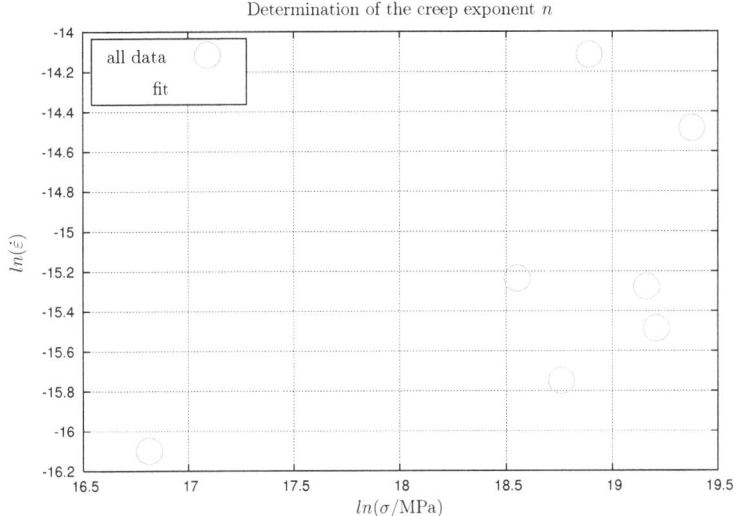

Figure 3.23: Strain rates versus stress to determine the creep exponent n. No clear correlation is visible.

In figure 3.24 the values obtained at room temperature have all been normalized to be 1. Some individual initial values for the HWHM at room temperature are given in table 3.10 and show that even for some thousands of fibres the initial values differ slightly, either due to the production process or due to the preparation. From the values listed in table table 3.10 the mean value at room temperature was calculated:

$$hwhm_{RT,mean} = (18.7 \pm 0.6)°$$

The first measurement at high temperature was usually performed directly after heating the fibre, which in in a time range between 5 min and 15 min.[14] A clear temperature dependence of the graphene layer orientation is visible. The higher the temperature the higher is the orientation (which means smaller HWHM). It is obvious that the main structural change occurs at the beginning of the creep experiment, during primary creep. After heating the fibre and after about 10 min this first drastic increase in orientation is finished. During secondary creep the orientation further increases slightly. A faster increase in orientation is visible for lower temperatures, slowing down

---
[14]See the creep curves as depicted in section 3.2.

3.3. Results

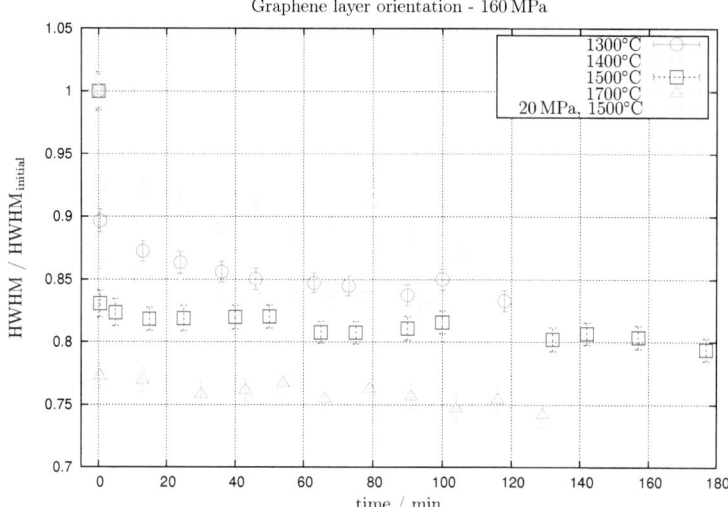

Figure 3.24: Mean graphene layer orientation determined by different experiments at 160 MPa. All measurements start at room temperature (time zero) - these values have been normalized to 1. The decrease of the HWHM with time is clearly visible. Higher temperatures show faster decrease. The final value for each data set is always obtained by a final measurement at room temperature.

Nr.	hwhm / °	Δ_{hwhm} / °
5	18.6	0.2
8	17.4	0.3
11	19.2	0.2
12	18.5	0.3
13	18.7	0.2
14	18.5	0.3
15	19.4	0.4
16	19.0	0.2

Table 3.10: Hwhm values of different carbon fibre bundle samples corresponding to the mean graphene layer orientation with respect to the fibre axis. Measurements have been performed on as received fibre bundles of the type HTA5131 at room temperature. The errors Δ_{hwhm} listed are the fit errors.

with time. In general, all fibres show a similar moderate increase in the orientation after 60 min.

3.3. Results

For comparison, the values of M16 (20 MPa, 1500°C) are also presented. This fibre bundle was tested at an stress below the creep threshold. A slight change in the graphene layer orientation was observed. The change is even smaller than for a temperature of 1300°C, but nevertheless clearly visible.
For all data series a final measurement was performed at room temperature at the end of the creep experiment. The corresponding final values are depicted at the end of each data series (highest time values) showing that the detected changes are permanent - the values do not return to the initial values at the beginning of the experiment.

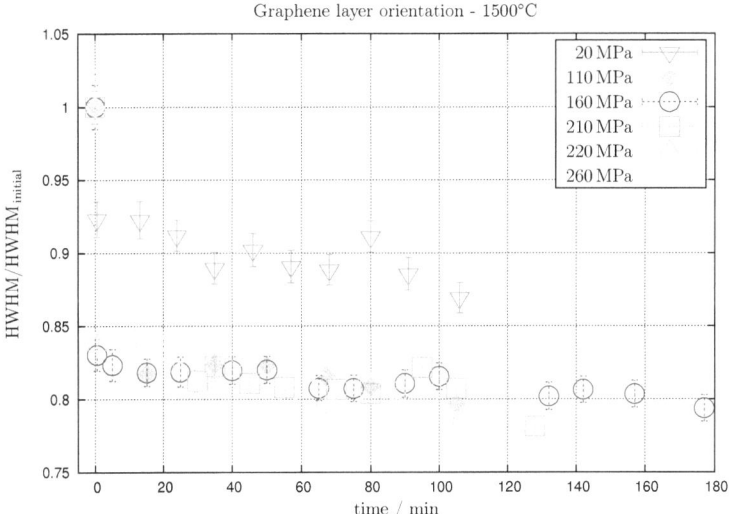

Figure 3.25: Mean graphene layer orientation determined by different experiments at about 1500°C. All measurements start at room temperature (time zero) - these values have been normalized to 1. The decrease of the hwhm with time is clearly visible. Except for the data of measurements at 20 MPa all samples show the same temporal evolution. The final value for each data set is always obtained by a final measurement at room temperature.

In figure 3.25 the room temperature values have also been normalized to 1, to visualize the relative change of the graphene layer orientation. The measurements have been performed at the same testing temperature of about 1500°C and at different stress levels. All samples show increasing orientation (which corresponds to a decreasing HWHM) with increasing time of the experiment. The main change in the HWHM is obtained in the first 10 min (during primary creep) while only a slight change can be observed

71

3.3. Results

afterwards (during secondary creep). In general, no dependence on the stress is observed, with the exception of the measurement at 20 MPa (M16), all samples show the same behaviour with respect to the error margin. The sample at 20 MPa shows a small decrease in the HWHM, which is about twice smaller as compared to the other samples, but which clearly indicates higher orientation of the graphene layers.

3.3.3 The inter-layer spacing

The temporal evolution of the inter layer spacing d_{002} during the creep process, calculated from the radial intensity distribution of the 002 reflection, is presented in figures 3.26 and 3.27 for different temperature values at constant stress and different stress at constant temperature, respectively.

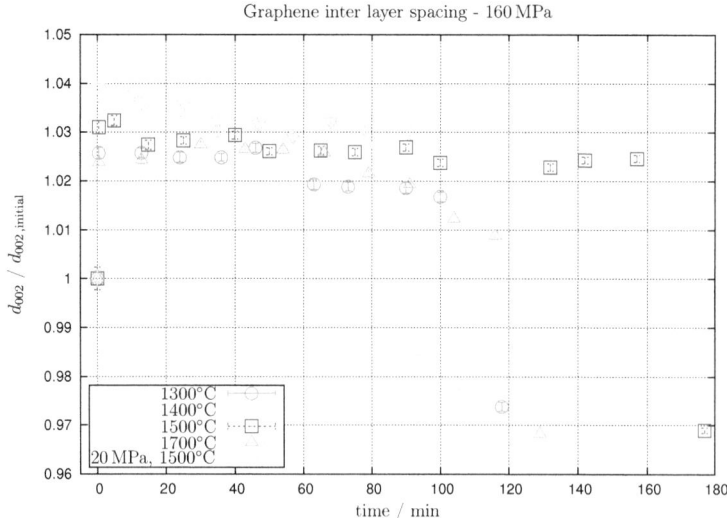

Figure 3.26: Graphene layer distance obtained by different experiments at about 160 MPa. All measurements start at room temperature (time zero) - the corresponding d_{002} have been norm normalized to 1. The inter-layer spacing is very sensitive to a temperature change. An increase of d_{002} at the beginning and a decrease at the end of the experiments, due to thermal expansion, is visible. A slight decrease of d_{002}, during the experiments is visible. No correlation between the maximum change and the test temperature is found. The final value for each data set is always obtained by a final measurement at room temperature.

3.3. Results

In figure 3.26 the values at room temperature measured before the experiments have been normalized to 1. The individual values obtained from evaluating the X-ray pattern can be found in table 3.11. A mean value has been calculated and was found to be:

$$d_{002,RT,mean} = (3.53 \pm 0.04)\,\text{Å}$$

The normalization of d_{002} is helpful for comparison of the results. A systematic error for the first value at testing temperature has to be taken into account, due to different detector positions relative to the sample. This leads to a systematic error in the data of about

$$\Delta_{system} = \pm 0.04\,\text{Å}$$

which does not inflict on the values within one data set. All fibres at first show an increase in d_{002} due to thermal expansion. For all data sets a slight decrease with increasing measuring time is visible. The final values for each of the data series is again obtained from a X-ray diffraction image taken at room temperature at the end of the experiment. All these values are smaller than the starting value, with a decrease between $0.06\,\text{Å}$ and $0.12\,\text{Å}$. There is no clear correlation between the testing temperature and the permanent decrease of the interlayer spacing.

At the end of M12 the temperature was decreased stepwise, leading to decreasing thermal expansion. The final value of M8 might be therefore affected by the considerably longer time period at high temperatures for this experiment.

Nr.	d_{002} / Å	Δ_{d002} /
5	3.590	0.008
8	3.510	0.008
11	3.50	0.06
12	3.584	0.006
13	3.538	0.004
14	3.495	0.004
15	3.508	0.006
16	3.500	0.006

Table 3.11: Interlayer spacing from the d_{002} reflection for different carbon fibre bundles. Measurements have been performed on as received fibre bundles of the type HTA5131 at room temperature. The errors Δ_{d002} listed are the fit errors.

In figure 3.27 the room temperature values at the beginning of the experiments have been normalized to be 1. In general, a slight decrease of

3.3. Results

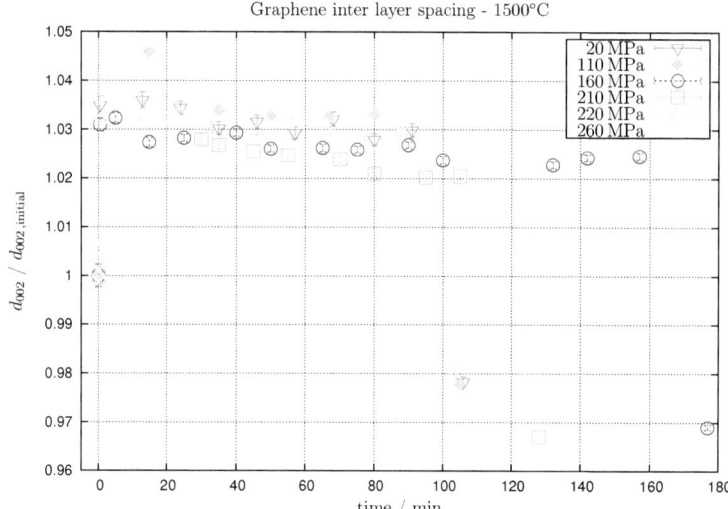

Figure 3.27: Graphene layer distance from different experiments at about 1500°C. All measurements start at room temperature (time zero) - these values have been normalized to 1. The values are sensitive to temperature changes. In general a decrease of d_{002} but no correlation between the maximal change and the test temperature is visible. The final value is always obtained by a final measurement at room temperature.

the inter layer spacing with testing time is visible. No correlation between the applied stress and the final permanent change in the inter layer spacing is detectable. The permanent decrease of d_{002} is found to be between 0.07 Å and 0.1 Å, while the decrease during creep seems to be only about 0.05 Å.

3.3.4 Radius of gyration

The radius of gyration was evaluated and plotted as a function of time comparing measurements at different temperatures but constant stress and different stress levels but constant temperature, respectively. The data are shown in figure 3.28 and figure 3.29.

The room temperature values at the begin of the experiment are normalized to 1, subsequently changing the corresponding values of each data set, as depicted in figure 3.28. The individual values of R_g measured before the experiment at the room temperature are listed in table 3.12. The mean value

3.3. Results

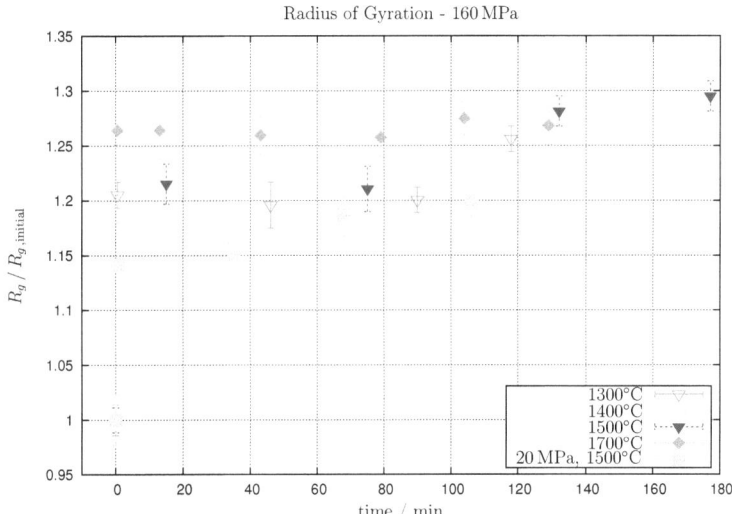

Figure 3.28: Radius of gyration for different temperatures at the same stress of about 160 MPa. All measurements at time zero are the room temperature measurements previous to the high temperature experiment - these values have been normalized to 1. An increase of R_g with test temperature is visible for all temperatures. The final value for each data set is always obtained by a final measurement at room temperature.

calculated is:

$$R_{g,RT,mean} = (0.44 \pm 0.01)\,\text{nm}.$$

The evolution of the radius of gyration with time during creep is comparable to the one of the HWHM as described before. In the first five to ten minutes (during primary creep) the value increases drastically. A correspondence of the primary increase of R_g with the test temperature is visible. With increasing testing time R_g increases. The sample at 20 MPa and 1500°C shows the smallest primary increase followed by a steady further increase during the measurement. The final value or each data set is measured again at room temperature and thus the change is obviously permanent.

The radius of gyration is permanently increased during creep. This is confirmed in figure 3.29.

All measurements have been performed at about 1500°C at different stress values. The radius of gyration is again found to be increasing during the creep experiment. In the first five to ten minutes a drastic increase is visible

75

3.3. Results

Nr.	R_g / nm	Δ_{Rg} / nm
5	0.429	0.006
8	0.437	0.005
11	0.44	0.01
12	0.46	0.01
13	0.436	0.006
14	0.42	0.01
15	0.459	0.08
16	0.445	0.009

Table 3.12: R_g values of different carbon fibre bundle samples corresponding to the interlayer spacing. Measurements have been performed with as received fibre bundles of the type HTA5131 at room temperature. The errors Δ_{Rg} listed are the fit errors.

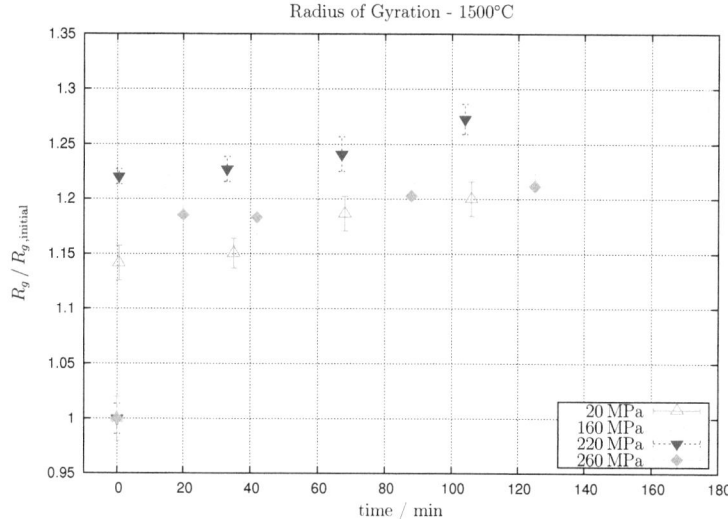

Figure 3.29: Radius of gyration of different experiments at about 1500 MPa. All measurements at time zero are the room temperature measurements previously to the high temperature experiment - these values have been normalized to 1. An increase of R_g during creep is clearly visible. The final values for each data series are always obtained by a final measurement at room temperature.

followed by a steady increase during the measurements. There is no clear correlation between the applied stress and the primary increase of R_g, which is also the case for the secondary creep.

3.3.5 Size of the graphene crystallite L_c

The size of the graphene crystallites in c-direction, e.e. more precisely the size of the coherently scattering graphene planes perpendicular to the planes, is presented in two graphs Figures 3.30 and 3.31 show the time evolution of L_c for different experimental temperatures at constant stress and for different stress values at constant temperatures, respectively.

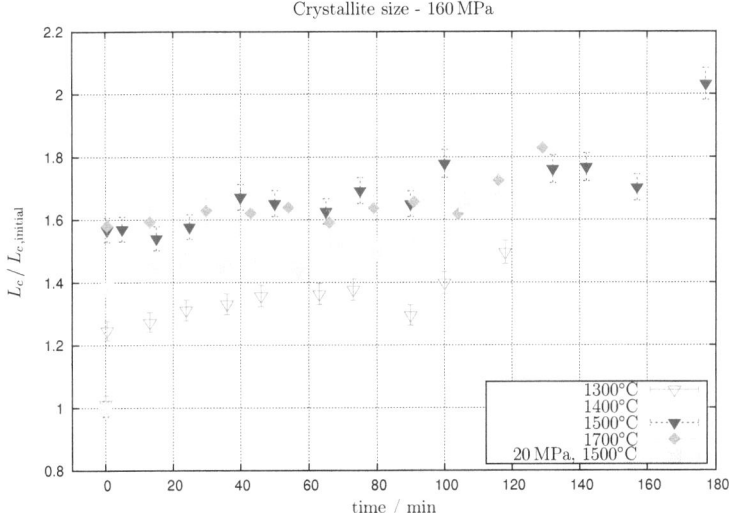

Figure 3.30: Crystallite size for different temperatures at the same stress of 160 MPa. All measurements start at room temperature (time zero) - these values have been normalized to 1. All samples show a clear increase of L_c with test temperature, which is also found to be temperature dependent. The final value of each data set is always obtained by a final measurement at room temperature.

In figure 3.30, the temporal evolution of the crystallite size L_c for different temperatures and constant stress level of 160 MPa is visible. The first values measured at room temperature have all been normalized to 1, to visualize the relative change of the crystallite size. The individual values obtained from the measurements at room temperature are summarized in table 3.13. The mean value was estimated to be:

$L_{c,RT,mean} = (1.6 \pm 0.1)\,\text{nm}.$

L_c is clearly increasing during the creep experiment. The main increase is

3.3. Results

Nr.	L_c / nm	Δ_{Lc} / nm
5	1.38	0.07
8	1.72	0.08
11	1.56	0.08
12	1.67	0.08
13	1.60	0.08
14	1.53	0.07
15	1.61	0.08
16	1.53	0.07

Table 3.13: Crystallite size L_c of different carbon fibre bundle samples. Measurements have been performed on as received fibre bundles of the type HTA5131 at room temperature. The errors Δ_{Lc} listed are the fit errors.

found within the first five to ten minutes, followed by a further but smaller increase with increasing time of the high temperature measurement. It is obvious that the higher the test temperature, the higher the crystallite size. The temperature effect is considerably stronger than the influence of the stress, as sample M16 (1500°C, 20 MPa) with a very low stress level exhibits an increase of the crystallite size, which corresponds to its test temperature.

In figure 3.31, the values of L_c are shown for the same temperature (1500 °C), but for different stress levels. All initial values have again been normalized to 1. A clear increase of L_c with increasing time of the experiment is found, but no clear correlation with the respective stress level is observed.

3.3.6 Thermal expansion coefficient

The thermal expansion coefficient was evaluated using the data of M12. After the creep process, the temperature was decreased to the values 1680 °C, 1580 °C, 1470 °C, 1300 °C, 1150 °C and room temperature (about 20 °C). At each temperature level, an X-ray pattern was taken and d_{002} was evaluated. The corresponding macroscopic elongation of the fibre bundle was measured by the LVDT. These data are presented in figure 3.32.

The values of the macroscopic expansion coefficient are normalized to zero at room temperature. The nominal value at room temperature was 680 μm which was simply subtracted. A linear fit is applied to each data set including values between 1000 °C and 1800 °C. The thermal expansion coefficients

3.3. Results

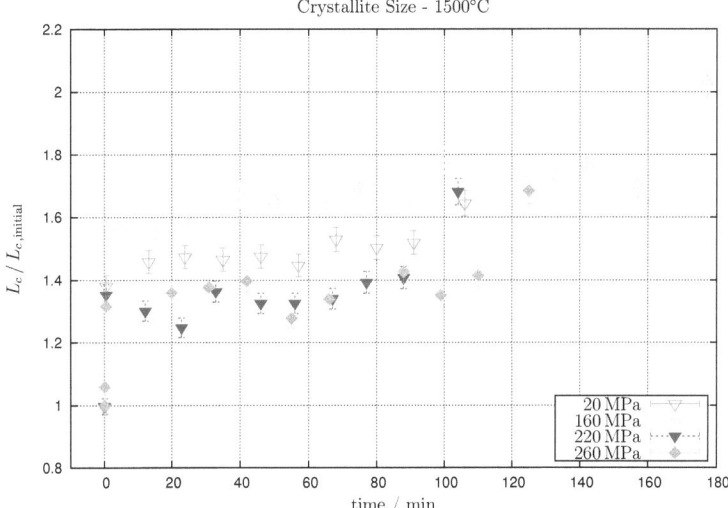

Figure 3.31: Crystallite size of different experiments at about 1500 °C. All measurements start at room temperature (time zero) - these values have been normalized to 1. An increase of L_c during creep is clearly visible. The final value for each data set is always obtained by a final measurement at room temperature.

are calculated using equation (2.1). The initial values for the gauge length and the interlayer distance are $l_{macroscopic} = 64$ mm and $l_{microscopic} = 3.47$ Å, respectively.

$$\alpha_{mac} = (1.83 \pm 0.09) \cdot 10^{-6}\, K^{-1}$$

and

$$\alpha_{mic} = (35 \pm 2) \cdot 10^{-6}\, K^{-1}.$$

From figure 3.32 it is observed that α_{mic} is nearly constant beginning at room temperature and up to 1800 °C, whereas α_{mac} is temperature dependent. The value of α_{mac} at a temperature of 1400 °C and 1700 °C is determined again by limiting the fit interval, respectively. The values thus obtained are: $1.0 \cdot 10^{-6}\, K^{-1}$ at 1400 °C and $2.2 \cdot 10^{-6}\, K^{-1}$ at 1700 °C.

It should be noted that these values should not be directly be compared: The microscopic value α_{mic} is corresponding to the out of plane direction and has to be compared with the radial expansion of a single carbon fibre, whereas the macroscopic value α_{mac} corresponds mainly to the in plane displacements of the atoms within the graphene sheets.

3.3. Results

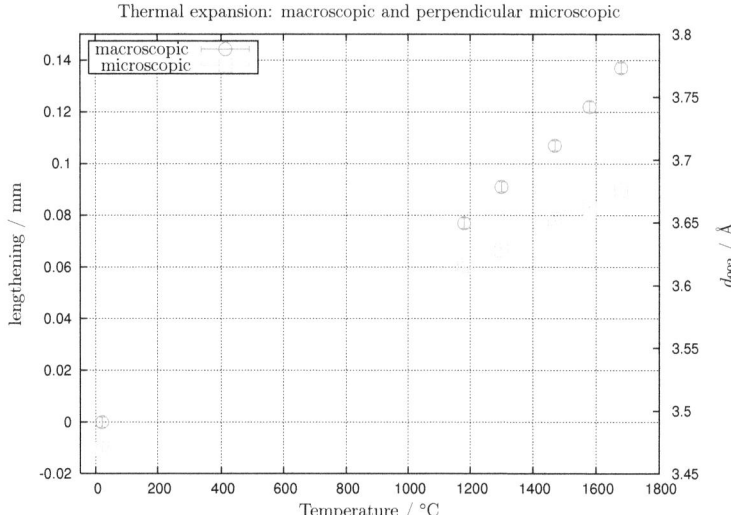

Figure 3.32: Data of the thermal expansion. The red circles indicate the macroscopic values of the expansion in fibre axis, referring to the left scale. The green squares indicate the microscopic values (d_{002}) of the expansion perpendicular to the fibre axis, referring to the right scale. The values of the macroscopic expansion have been normalized to zero at room temperature.

3.3.7 Chemical analysis

After the creep experiments (heat treatment with additionally applying stress) the carbon fibre bundles were prepared for chemical analysis. The fibres were cut into small pieces and micro-analytical elemental analysis (EA 1108 CHNS analyzer from Carol Erba instruments) was used [8] to determine the content of carbon, hydrogen, nitrogen and sulphur. The samples were compared to as received fibres and fibres heat treated without load. The sample material was always taken from the middle of the gauge length of the carbon fibre bundle. The results which were already published in part by the author of this thesis in [50] to interpret the growth of the crystallites in PAN-based carbon fibres, are summarized in table 3.14. A decrease in the nitrogen content from 4.1 % for the as received fibre to less than 1.4 % for the fibres treated at 1500 °C under load and even smaller than 0.1 % for fibres treated at more than 1800 °C is found. The carbon content is increasing with increasing temperature. No hydrogen is detected for any heat treated fibre within the error margin.

3.3. Results

sample	conditions	C / $w-\%$	H / $w-\%$	N / $w-\%$
USG 001	as received	94.7	0.2	4.1
USG 003	1500 °C, 20 MPa	97.9	< 0.05	1.4
USG 004	1500 °C, 220 MPa	98.6	< 0.05	0.84
USG 005	HT 1800 °C	98.3	< 0.05	0.08
USG 002	HT 2100 °C	99.1	< 0.05	< 0.05
USG 006	HT 2400 °C	98.7	< 0.05	0.11

Table 3.14: Results of the chemical CHNS analysis, where $w-\%$ stands for percent by weight. All errors are of the dimension of one digit in the last digit presented. Samples 003 and 004 were used in an creep experiment. All values for S were smaller than 0.02 $w-\%$.

Comparing the samples 003 and 004, which were subjected to the same temperature but to different stresses, a higher amount of carbon and less nitrogen was observed for the sample with the higher stress. This leads to the conclusion that the structural change takes place at certain temperatures, but is additionally enhanced by the applied stress.

Chapter 4

Experiments with single carbon fibres

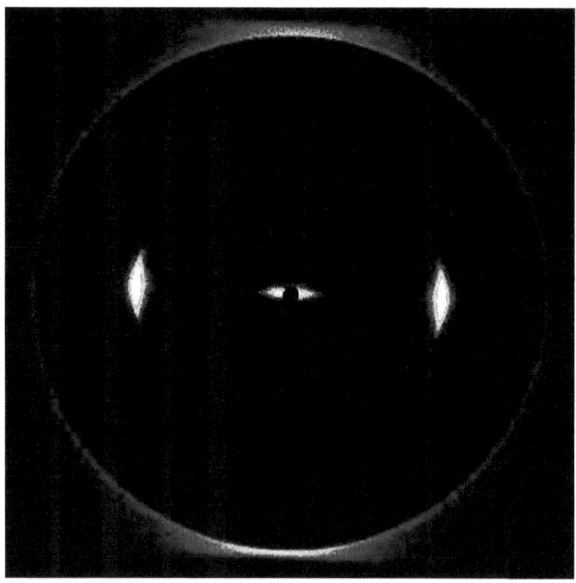

Figure 4.1: A characteristic pattern taken by X-ray diffraction of a K137 single fibre. The ex-pitch fibre K137 shows sharp reflections, furthermore the 10 band is clearly visible.

Single carbon fibres were tested because in the fibre bundle it is impossible to separate the misalignment from fibres in a bundle from the misalignment of the graphene sheets within the single fibre. The mean angle referring to the misalignment of the graphene sheets or pores, respectively, is higher if measured for carbon fibre bundles. A carbon fibre bundle consists of thousands of single fibres and thus all obtained results represent an average of an ensemble of individual fibres, which have individual properties. These properties, however, might vary slightly, from fibre to fibre. Single fibre tests have particular advantages with respect to a significantly higher precision in the determination of the structural parameters from X-ray diffraction. On the other hand experiments are much more challenging, as the diameter of one single fibre is only about $7\,\mu$m. This small scattering volume makes the use of the high intensity of the X-ray beam in a synchrotron radiation source inevitable. Furthermore, a micro beam with a beam width in the size of a fibre is required to prevent from high parasitic background scattering (made e.g. from the X-ray windows) and to obtain a signal to noise ratio as high as possible.

For the single fibre tests a special testing device was constructed. The setup (section 4.1) as well as the measuring procedure (section 4.2.5) were quite different (as described below) from that applied for carbon fibre bundles (sections 3.1 and 3.2.3). Experiments have been carried out in a vacuum vessel at pressures well below 10^{-4} mbar. The single fibres are directly heated by the electrical current. The temperature of the fibres is measured by the heating power, which is calibrated before the experiments (section 4.2.4). Due to the small heating power no heat shield is required and thus the setup allows to record not only the 002 reflection but also the 10 band. The setup has the advantage of simplicity in the construction, but the disadvantage that no displacement curve is recorded because measurements are done by the dead weight method.

For accurate measurements the precise fibre temperature and the stress applied to the fibre have to be known. Therefore the calibration of the temperature measurement has to be performed with utmost care. Further, knowledge of the gauge length of the fibre, its precise diameter and the weight applied to the fibre is crucial. The first is measured before the experiment, the latter two are estimated in first step from calibrations based on the fibre resistivity and the dimension of the Sigraflex foil. They are precisely determined after the experiments in a SEM and by a high precision balance, respectively.

4.1 Experimental Setup

The setup for the single carbon fibre experiments was specially designed for the application in a synchrotron radiation source. It consists of a small vacuum vessel and a sample holder, see figure 4.2.

The vessel has three flanges, one to attach the vacuum pump, one with

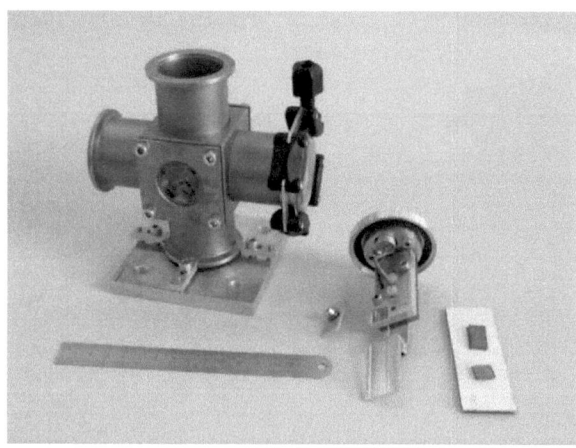

Figure 4.2: Single fibre setup. The vessel, the sample holder and a typical prepared single fibre are depicted. The Kapton foil and the glass window sealing the vessel are visible. The flange connected to the pumps and the top flange, for the sample holder are open.

a glass window that the fibre is visible and the top flange to mount the fibre (to insert the sample holder). Two further flanges are covered by foils transparent for X-rays. The X-ray windows are made from *Kapton* at the detector side and *Kalle Bratfolie* at the side the beam enters the vessel. *Kalle Bratfolie* is chosen to minimize the background scattering, because it has less background scattering than *Kapton*. However, its mechanical strength is too poor to sustain the air pressure for the larger flange. Thus, it is only used for the small window, where the X-ray beam enters the vessel. The bottom part of the vessel is isolated with silicon foil. This prevents from a short circuit, which would damage the heat control, if the dead weight touches the bottom after fibre failure.

The sample holder is built from aluminium and consists of two isolated parts, which are also used as power lines, as shown in the scheme in figure 4.3.

4.1. Experimental Setup

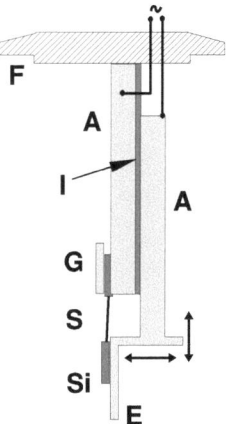

Figure 4.3: Scheme of the single fibre setup. F...flange, A...aluminium part, I...insulating part, G...grip, S...single carbon fibre, Si...Sigraflex sandwiching the fibre, E...extension movable in horizontal and vertical direction. The power supply is indicated by two lines. Screws or connecting parts are not shown.

One of them is attached to a standard ISO-KF40 flange which has a *LEMO* feed through for the power lines. The single fibre is fixed to this part during the experiment. The other aluminium plate is mounted to the plastic isolation with non conductive plastic screws. It has an extension to contact the lower end of the fibre and completes the current circuit. The position of the sample contact of the extension can be adjusted for different gauge lengths by moving the aluminium plate vertically or the extension horizontally, to ensure good electrical contact. The vessel is designed in such a way that photons scattered up to 30 degrees (for Cu-K$_\alpha$) or a scattering vector of $q = 4\,\text{nm}^{-1}$ pass the X-ray window of the vessel on the detector side and hit the detector. The diameter of this window is 30 mm.

The vessel itself is mounted on an xyz-stage. The base of the vessel has to be attached very firmly to the experimental table, to suppress vibrations from the pumping system. These vibrations would lead to a loose contact between fibre and aluminium plate resulting in a non-uniform temperature. The direct beam is stopped between vessel an detector with a beam-stop consisting of a 0.1 mm lead wire. The end of the wire directing to the beam is formed like a cup to avoid parasitic scattering and total reflections. The wire is glued to a bent glass capillary and is adjusted by an xyz drive. A photo of the vessel with beam-stop in the experimental hutch of the micro-

4.1. Experimental Setup

focus beamline of BESSY is shown in figure 4.4.

Figure 4.4: Setup of the experiment. B...beam-stop, M...microscope for the fibre adjustment (details in section 4.2), V...connection to the vacuum stage (green tube), P...plug contact for the power supply.

Each end of a single fibre is glued with electrical conductive adhesive (conductive Epoxy CW2400 from *circuitworks* containing silver particles) between two *Sigraflex* platelets. Curing is achieved in 4 hours at room temperature. One end of the fibres is then gripped and thus fixed to one of the aluminium plates. The other end of the fibre is used for the dead weight, contacting the extension of the second aluminium plate, see figure 4.3.

For the preparation of a sample, a part of a carbon fibre bundle is usually put into a white dish filled with water. The white colour helps that the single fibres are visible with the naked eye, the surface tension of water helps to separate single fibres from the bundle. Sigraflex platelets are prepared for the top part with dimension of $1\,\text{cm}^2$. For the bottom part, which acts as dead weight, the width of the platelets is 1 cm, but the length is adjusted to the specific weight required for a certain nominal stress acting on the fibre. The Sigraflex platelets are covered with the epoxy adhesive and the fibre is placed in the middle of the platelets. Adhesive is also applied to the covering Sigraflex parts.

The whole procedure has to be performed very carefully, as the single fibres easily fracture.

4.2 The experiments

4.2.1 Conducting adhesive

Different adhesives have been intensively tested to find the best glue for the single fibre experiments. The adhesive should have good contact to the Sigraflex and the carbon fibre. It should be cured at room temperature for the reason of simplicity in sample preparation. Furthermore, it should not destroy the fibre due to chemical reactions.

In a first approach we used a two component epoxy adhesive mixed with graphite powder. The maximum amount of carbon powder, which could be added to the single components before mixing them was 50 % of the total end mass. A higher amount of powder leads to a dry mixture, which had not enough viscosity to give a useful glue. The procedure was to add the graphite powder to the components of the glue during continuous stirring. Heating to 50 °C to increase the viscosity did increase the amount of powder, which could be mixed into the adhesive, but not to a degree, which would allow uniform and reproducible properties with respect to shear strength and conductivity. The glue and the curing agent of the adhesive, each mixed with graphite powder, can be kept for more than a month, which has the advantage of fast sample preparation at the synchrotron, but the disadvantage of insufficient conductivity. Bad contact within the adhesive and from the graphite powder to the comparably small fibre might be the reason. Even the use of short carbon whiskers mixed together with the adhesive was not improving the conductivity.

The graphite adhesive *Respond* 931P graphite putty was tested, which cures at room temperature as well as at 100 °C and which contains of 90 % pure carbon after curing. This glue showed surprisingly bad results, concerning an insufficient adhesion of the glue on the Sigraflex platelets.

Finally two different epoxy adhesives were tested: the RS186-3616 silver loaded epoxy adhesive from *RS-Components* and the two-component epoxy adhesive cw2400 (also with silver load) from *circuitworks*. The latter showed good results together with low costs and was taken as standard adhesive for the experiments.

4.2.2 Calibrating the stress

Nominal stress values expected from a certain dimension of dead weight (including the adhesive) have been calibrated and are used during the experiment. The exact stress values are determined after the experi-

4.2. The experiments

ments by weighting the Sigraflex compound. Using the real diameter of the fibres, also determined post creep, an exact stress value can be calculated.

The mass of the Sigraflex - adhesive - Sigraflex compound was calibrated by taking the mean value from twelve samples. The preparation method, in particular the amount of adhesive, was identical to the one used in the in-situ experiments. Eight samples were of dimensions 1 cm × 4 cm and four samples of dimensions 1 cm × 1 cm. Each single mass was measured with an error of 1 mg and the mean value of the mass for a piece of compound of the dimension 1 cm × 1 cm was calculated with its error. This reference mass $m_{sig,r}$ is:

$$m_{sig,r} = (310 \pm 40)\,\text{mg}$$

Taking 7 µm as a mean diameter of the carbon fibre, the mass corresponding to a stress of 217 MPa was calculated using $g = 9.81$ ms^{-2} for the acceleration of gravity:

$$m_{217} = 851.3\,\text{mg}$$

From these two values, the nominal masses for the dead weight method and the subsequent dimension of the Sigraflex compound are calculated. Three nominal stresses have already been applied in the fibre bundle tests and for the reason of comparison are chosen again for the single fibre tests. The corresponding mass and length of a Sigraflex platelet are given in table 4.1. In addition the mean value for the mass of one single Sigraflex platelet was

σ / MPa	m / mg	$L \pm \Delta_L$ / mm
173	678.7	22 ± 2
217	851.3	27 ± 4
260	1020.0	33 ± 4

Table 4.1: Calibration of the Sigraflex compound mass: σ gives the nominal stress, m the corresponding mass, and L the corresponding length of a Sigraflex compound (Sigraflex - adhesive - Sigraflex) having the width of 1 cm together with its error Δ_L.

estimated using another 11 samples. The mass of the 1 cm × 1 cm Sigraflex platelet is $m_{sig} = (110 \pm 5)\,\text{mg}$ and the mean mass of the adhesive to cover a square centimeter of the compound is: $m_{adhesive} = (90 \pm 10)\,\text{mg}$.

4.2.3 Single fibre resistivity

The electrical resistivity of a single carbon fibre was determined via the linear relation between fibre length and resistivity. This resistivity was used in the calibration process of the fibre temperature as well as in the fibre bundle tests to estimate the fibre number as described in section 3.2.2. Measuring the electrical resistivity by contacting the Sigraflex compounds does not allow the separation of the resistivity of the glue and Sigraflex platelets from the resistivity of the sample. The resistivity of glue and Sigraflex was measured directly to be $100\,\Omega$ at maximum, but the experiment shows higher resistivity values, originating from the contact of the fibre to the glue. Samples of different length were prepared and the resistivity of these was measured. Measurements were performed for all adhesive types used, but only the data for the cw2400, which was further used for the in-situ experiments, are presented. Figure 4.5 shows one example out of many measurements for the resistivity of the Sigraflex-glue-fibre compound in dependence on the gauge length of the fibre. A linear dependence $R(l)$ is observed, and a linear regression allows therefore the calculation of the residual resistivity R_0 of the Sigraflex-glue compound from the limit to zero fibre length. With the linear function $R(l) = a \cdot l + R_0$ the following values were obtained for the adhesive cw2400:

$$a = (423 \pm 16)\,\Omega\,\text{mm}^{-1} \qquad R_0 = (1.1 \pm 0.2)\,\text{k}\Omega$$

The resistivity in the case of the RS silver-epoxide adhesive was similar: $a = (435 \pm 30)\,\Omega\,\text{mm}^{-1}$ and $R_0 = (1.1 \pm 0.3)\,\text{k}\Omega$. The value for the residual resistivity R_0 is surprisingly high and is therefore taken into account in the calibration of the temperature from resistivity measurements.

An error in the diameter of the fibres of about 5 % leads to an approximated error of approximately 10 % in the resistivity. Despite that for fibres with small diameter the resistance is higher and the slope of R(l) is increasing with similar residual resistivity R_0, the deviation from the linear fit is much lower, if fibres with similar diameter are grouped together as shown for the data set in figure 4.5.

4.2.4 Calibration of the temperature measurement

The fibre temperature is determined by measuring the heating power and calculating the temperature using the well known Stephan-Boltzmann relation, equation (4.1). When $P = U \cdot I$ is the electrical power calculated with

4.2. The experiments

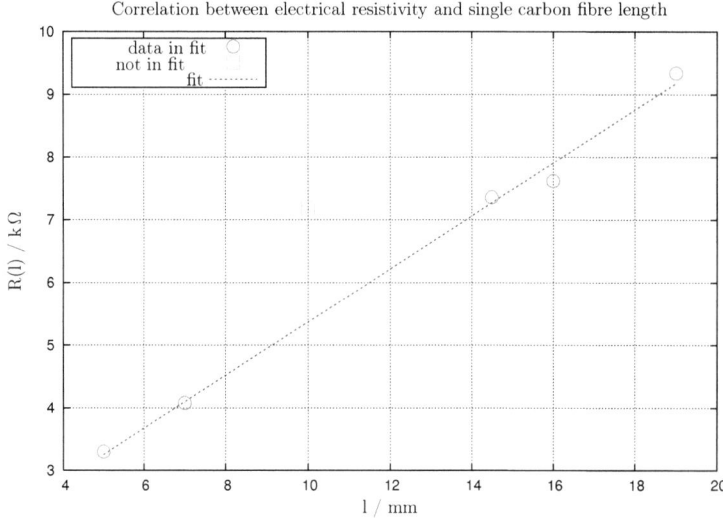

Figure 4.5: Calibration of the resistivity of a single carbon fibre of the type HTA5131. There is a clear correlation between the length l and the resistivity $R(l)$. The value indicated with the green square is only included in the figure to demonstrate the influence of a different fibre diameter. It corresponds to an ensemble with a different residual resistivity R_0. In this case the diameter would have been 16 % smaller than the mean value.

the voltage U and the current I, σ is the Stephan-Boltzmann constant[1], ε is the emission coefficient, A is the surface area and T is the temperature, then in general the following equation holds:

$$P = \sigma \cdot \varepsilon \cdot A \cdot T^4 \tag{4.1}$$

Both, current and voltage can be measured during the experiment using two $FLUKE$ multi-meters in the power line. The gauge length is measured and a computer program written by the author calculates the temperature and its error. The calculation is performed with a fibre diameter that is estimated from the resistivity of the fibre. The true fibre diameter is determined after the experiment and good correspondence of the estimated values is found (table 4.2). For the emission coefficient the value found for fibre bundles (section 3.1.6) was taken:

$$\varepsilon = 0.8$$

---

[1]The value of sigma is: $\sigma = 5.67 \cdot 10^{-8}\,\text{Wm}^{-2}\text{K}^{-4}$, taken from [4].

4.2. The experiments

and was found to be reliable during the calibration procedure, because the experiments showed good agreement with the theory applying this value.

The basic assumption of the calibration of the heating power is that nearly all of the power is irradiated by the fibre to the environment - only negligible amounts of power are lost in the power lines, the Sigraflex or the glue as heat. Thus, in principle equation (4.1) is valid for the description of the relation of temperature of the fibre and electrical power applied. The directly heated bright fibre in front of an also bright area with known temperature (comparable to a reference pyrometer) is used to calibrate the heating power. Crucial to this method is that the slightest brightness differences between the fibre and the background should be distinguishable, at least to guarantee a minimum resolution of the method. The human eye is very sensitive to differences in brightness, and grey filters are used to optimize the brightness to the human eye. The preparation procedure, including the Sigraflex platelets and the conductive adhesive, is now identical to the one used in the in-situ single fibre tests. The third critical point, which has to be addressed is the background temperature. As this is the reference temperature, it has to be measured with the highest precision possible.

The setup for the calibration measurements

The calibration experiments are performed in a vacuum vessel at second stage vacuum. As background a C-C composite rod heated by a graphite susceptor is used. Thus, the basic setup for the background is similar to the one in section 3.1.6 for the calibration of the pyrometer (see also figure 3.9). The C-C rod is fixed in the susceptor and the thermocouple monitors its core temperature. The bichromatic pyrometer measures the temperature of the surface of the rod through a hole in the susceptor. The setup is completed by the single fibre, which is mounted into the optical path of the pyrometer as shown in figure 4.6. The single fibre is prepared identical to an creep experiment and clamped with iron grips. Power lines run from a regulation transformer to the grips and thus the fibre can be heated to different temperatures. The current is measured with a *Keythley* multi-meter and the voltage with a *FLUKE* voltmeter. The error of the current measurement is 0.05 mA and the one of the voltage measurement 0.1 V, which corresponds to about 1 % for the current and 0.5 % for the voltage.
For the measurements the C-C rod in the susceptor is heated to a certain constant temperature which is measured with the pyrometer and the thermo couple. The carbon fibre is heated until it is brighter than the

4.2. The experiments

Figure 4.6: The setup to calibrate the single fibre temperature. In the background the graphite susceptor electrically isolated from the induction coil is visible with its front hole and the thermo couple entering from the left. In the foreground two grips are holding the Sigraflex compounds the fibre is glued into - which is too small to be visible in the photo. The power lines and the heat shields to prevent the grips from getting too hot are also visible. On the right side the glowing fibre is visible in front of the cold susceptor.

C-C background. Then the power is reduced until the fibre can just not be distinguished from the background. During the measurement two transmission filters are used, one for low temperatures and one for high temperatures, respectively.[2] The current and the voltage are recorded.

To increase the accuracy of the measurements, several strategies were used: The heating of the single fibre was repeated with a small pause in-between to relax the eye. An upper limit for the measured power was found by first drastically reducing the power and then increasing the power until the fibre was just visible. This value obtained for the power is slightly higher than the one achieved for a fibre, which disappears in the background of the heated C-C rod. Frequently the measurements were performed using both filters for comparison, or by two individual experimentalists to exclude systematic errors. It turned out that the power required to heat the fibre to a certain temperature was perfectly reproducible in all the above mentioned cases. The error in voltage and current were about 1 %. The gauge length of each sample was measured before the calibration with an accuracy of 0.1 mm.

The calibration on the basis of the Stefan-Boltzmann law

For each fibre the data for the heating power in dependence on the temperature $p(T)$ were normalized with respect to the gauge length of the fibre. In figure 4.7, all measured data are shown. All fibres give nearly the same

---

[2]The precise temperature at which the filters were changed differs from experiment to experiment. Usually it is between 1300 °C to 1400 °C.

behaviour of the normalized heating power $p_{norm}(T)$ with the exception of fibre C6. This is probably due to the exceptional small diameter of the fibre, which led to a poor visibility, and it was extremely difficult to distinguish this fibre from the background, the heated C-C rod. Therefore this fibre is not used in the following fit procedures.

The theoretical curve from the Stefan-Boltzmann law, equation (4.1), is shown with the index *norm* in figure 4.7. The numerical values are derived by inserting a specific fibre diameter d_0 and a specific normalized gauge length l_0:

$$d_0 = 7.0\,\mu m \qquad l_0 = 15\,mm$$

It is visible from figure 4.7 that the trend in the experimental data and the theoretical curve is the same, the only difference is, that the measured values are shifted vertically to slightly higher values. The reason being is a contact resistance between the grips (i.e. the Sigraflex platelets and the glue) to the fibre. Thus, equation (4.1) is supplemented by a term, which takes a finite contact resistance into account. This parameter h just adds to the Stefan-Boltzmann law and has a dimension of the normalized heating power.

When $p_{norm,h}(T)$ is the normalized power, which takes the contact resistance into account, σ the Stephan-Boltzmann constant, ε the emission coefficient, which numerical value $\varepsilon = 0.8$ was determined previously in section 3.1.6, $A = d\pi l$ the surface of the fibre (with d the fibre diameter and l its length), T the fibre temperature and T_0 the temperature of the environment, h is the mean deviation of all data sets from the theoretical curve, then the following equation holds:

$$p_{norm,h}(T) = \frac{\sigma \cdot \varepsilon \cdot A \cdot (T^4 - T_0^4)}{l_0} + h \qquad (4.2)$$

From a fit, the vertical shift by the contact resistance induced parameter h is obtained:

$$h = (0.84 \pm 0.05)\,mW\,mm^{-1}$$

For the fit d and l were taken to be d_0 and l_0. T_0 can be neglected, as it is small in comparison to T^4 for the high temperatures in our experiment. The fit curve is shown with the index *fit* in figure 4.7.

This procedure copes with the effect of the contact resistance of the grips, which is mainly the resistance between glue and fibre. It is, however, based on a constant nominal fibre diameter of 7.0 μm and does not take the statistical distribution of fibre diameters into account. However, a variation of the fibre diameter d to describe the data set for each fibre separately, allows a fine tuning of the calibration method, which is described in the next section.

4.2. The experiments

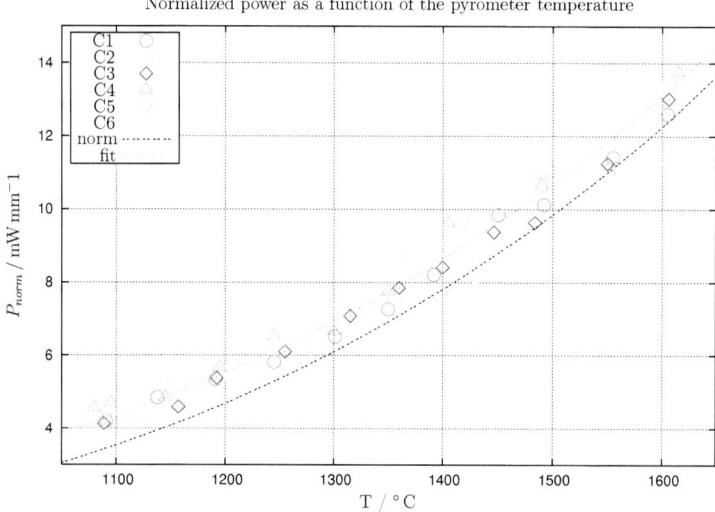

Figure 4.7: Calibration of the heating power of the single carbon fibres. A fit was applied to all data except the data of C6. The data points from the measurements are shown together with the normalized curve $p_{norm}(T)$ with the index *norm* and the curve adjusted with the parameter h to all data by a fit, with the index *fit*. All data sets show the same trend as the theoretical curve (derived from the Stefan-Boltzmann law) with small systematic deviations due to different fibre diameters and small statistical deviations due to a change in the fibre temperature during the measurements.

Fine tuning of the calibration procedure

With the given setup (for the experiments at the synchrotron radiation source), it is only possible to determine the electrical resistivity in dependence on temperature and the gauge length of the fibres. The estimation of a hypothetical fibre diameter D_f from these two measurements significantly enhances the quality of the temperature determination of the single fibre in the in-situ experiment. Therefore a correlation between D_f and the electrical resistivity of the single carbon fibre is determined:
The influence of the contact resistance of the Sigraflex-glue system was taken into account by equation (4.2) and experimentally determined from the fit in figure 4.7 to all data values. The only remaining unknown variable is the fibre diameter. A variation of the fibre diameter d to describe the data set for each fibre separately leads to hypothetical diameter values D_f. An example of the fit is shown in figure 4.8. To verify the validity of this procedure, all

the fibres were kept and investigated in the SEM after the calibration procedure, from which the actual diameter D_m was obtained. A comparison of the diameters obtained from both methods, the fine tuning of the function $p_{norm,h}(T)$ via the variation of the parameter d and the direct determination in the SEM, respectively, is presented in table 4.2. Both values are in nearly perfect coincidence.

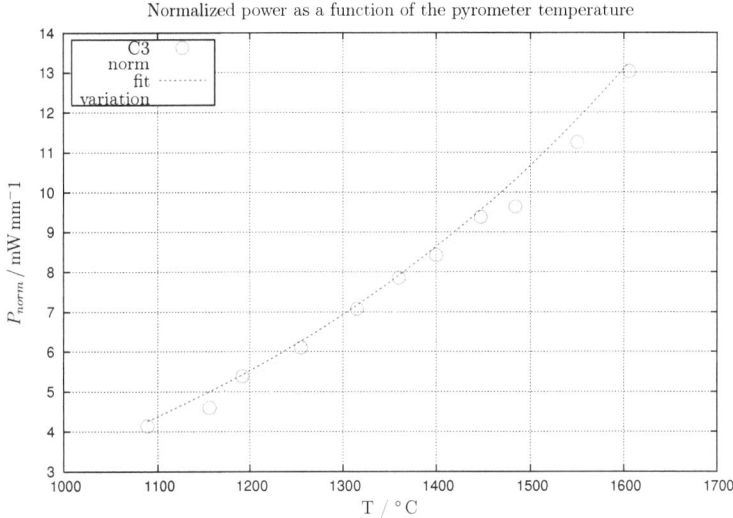

Figure 4.8: Variation of the fibre diameter d to describe the data set of measurement C3. The fit functions $p_{norm}(T)$ with the index $norm$, $p_{norm,h}(T)$ (with the diameter d_0) with the index fit and the same function after variation of the fibre diameter with the index $variation$ are shown in the graph. $p_{norm,h}(T)$ is obtained by a fit of all data sets with the parameter h. Thus, the data of fibre C3 and the curve after variation of the fibre diameter deviate from this mean value, i.e. are found at smaller values, which corresponds to a smaller fibre diameter.

The calibration procedure runs now as follows: The resistivity of each of the carbon fibres (C1 - C6) was measured before, during and after the temperature calibration. At the beginning the resistivity at room temperature is determined by application of low voltage and current, which is sufficiently small not to heat up the fibre significantly. During the calibration, the temperature and the corresponding resistivity of the fibre were measured stepwise at temperatures in intervals of approximately 50 °C. Finally, the resistivity was measured at room temperature, which might differ to the first measure-

4.2. The experiments

ment due to a permanent structural change of the fibre. This was observed for the sample C5, which was only measured after heating already once. Due to the change in resistivity after heat treatment, for this fibre also the diameter D_f differed considerably from the diameter D_m. Therefore, this fibre was not taken into account for further calculations.
The relevant data of the measurements are listed in table 4.2.

Nr. / #	R_0 / kΩ	L / mm	$\frac{R_0}{L}$ / $\Omega\,\text{mm}^{-1}$	D_f/μm	D_m/μm
C1	7.84	16.3	482	6.73	6.65
C2	7.11	16.2	438	7.05	7.01
C3	6.67	13.4	497	6.75	6.79
C4	7.08	16.4	431	7.23	6.89
C5	(7.20)	14.8	(486)	(7.46)	6.43
C6	11.11	20.2	551	5.83	6.16

Table 4.2: Data of the single fibres used for calibration of the temperature measurement. R_0...resistivity determined via voltage and current before the measurement with an error of about 0.05 kΩ, L...length of the fibre with an error of 0.1 mm, $\frac{R_0}{L}$...normalized resistivity with an error of about 5 Ω, D_f...hypothetical diameter determined by fit with an error of about 0.07 μm from the fit - see text for details, D_m...diameter determined after calibration by SEM with an error of 0.05 μm.

The relation between the normalized resistivity $\frac{R_0}{L}$ and the fibre diameters D_f is given by $R \sim \frac{L}{d^2 \pi}$. If a is a constant, d is the fibre diameter and L its length, D_f can be expressed:

$$D_f\left(\frac{R_0}{L}\right) \sim \sqrt{a \cdot \left(\frac{R_0}{L}\right)^{-1}} \qquad (4.3)$$

A fit for the function is shown in figure 4.9 together with experimental data. The fit parameter obtained is:

$a = (21500 \pm 700)\,\Omega\,\text{mm}^{-1}$

In the experiment at the synchrotron the gauge length and the resistivity of the fibre are measured. The resistivity is determined by applying a small voltage to the fibre and measuring voltage and current. The relation $D_f\left(\frac{R_0}{L}\right)$ is used do determine a hypothetical diameter. This diameter is used in equation (4.2) to correlate the fibre temperature and the heating power applied to the fibre.

4.2. The experiments

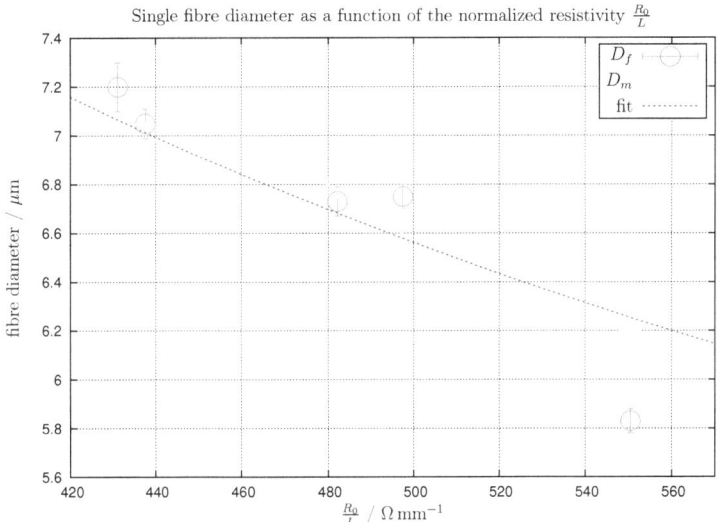

Figure 4.9: Fit for the relation between D_f and $\frac{R_0}{L}$. The square root function is shown together with the corresponding data D_f. For comparison the data of D_m obtained from SEM measurements are also shown.

A search for alternative optical measurement methods to determine the fibre temperature was performed, but no company could be found, which produces pyrometers with a high resolution optics to reliably measure objects less than 20 µm. Thus, the temperature calibration as described in the above section was really challenging and the reliable determination of the temperature of the fibre from its resistivity was a step towards realizing in-situ experiments at high temperatures and high loads. The method further allows an estimation of the maximum error taking all individual error contributions into account.

The error of the temperatures

For the estimation of the error, *Gaussian* error calculus was used and the maximum individual contributions of error were taken into account. The main contributions arise from the error in the gauge length and the error from current and voltage, whereas the fit error is comparably small. This is an very conservative estimation of the error and the actual difference of measured and calculated temperature is supposed to be much lower, as the maximum error of the measurement instruments was inserted into

4.2. The experiments

the numerical calculation. As an example, the error calculated for the two samples C12 (a HT1800 fibre) and M5 (a HTA5131 fibre) is shown in figure 4.10. Due to higher precision in the determination of the current, the error decreases for higher temperatures. An error of about 30 °C at a temperature of 1600 °C is very satisfactory. Thus the error in the

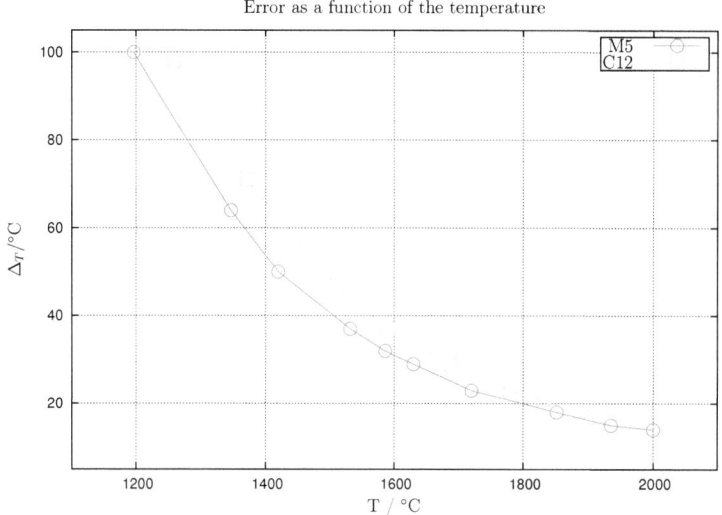

Figure 4.10: Calculation of the maximum error of the single fibre temperature measurements as a function of the temperature. The error is presented for two representative samples, the HT1800 fibre M5 and the HTA5131 fibre C12, respectively. Individual values have been calculated based current and voltage measured and are connected with straight lines. Values at temperatures higher than 1850°C are achieved by extrapolation of the given data.

temperature can be estimated with about 4 %, but will usually be smaller for temperatures higher than 1400 °C.

A final remark concerns the sizing of the carbon fibres, which is important for the determination of the diameter. The as-received HTA5131 carbon fibres are delivered by the company with sizing. The diameter of a HTA7 fibre with sizing is $(6.91 \pm 0.02)\,\mu m$ [51] and that of the same fibre without sizing $(6.78 \pm 0.02)\,\mu m$ [48]. Thus a thickness of the polymer sizing of about $(0.065 \pm 0.015)\,\mu m$ is derived. A similar value has to be assumed also for the HTA5131 fibres. The influence of the sizing with respect to the fibre strength is described in [37].

For our experiment, the important point is that the sizing is not electrically conductive and could thus affect the contact between fibre and adhesive, whereas it does not affect the fibre diameter with respect to the resistivity of the fibre. The polymer is burnt off at temperatures higher than about 400 °C and thus the diameter without sizing has to be taken into account for the temperature calculations. In the SEM, the total diameter - carbon fibre together with sizing - is observed. Therefore, we decided not to measure the fibres in the SEM in advance, but after the experiment. Only in this case the fibres, now consisting purely of carbon, are observed.

4.2.5 Measurement procedure

The single fibre experiments were all performed at the *BESSY* μ-Spot beamline in Berlin. A micro-focus of about 15μm to 20μm beam diameter at the fibre position was achieved by a Bragg-Fresnel lens. The wavelength of the beam after the double-crystal monochromator was

$$\lambda = 0.8266 \text{ Å}$$

which corresponds to an energy of 15 keV. A MarMosaic 225 CCD detector with a number of 3072 × 3072 pixels and a pixel-size of 73 μm was used. Binning of the pixels was applied, which reduces the number of pixels to 1536 × 1536.

Two main measurement procedures were chosen: One is to heat the fibre to a certain temperature and to record successive X-ray patterns at constant temperature. This is precisely the same procedure as the corresponding measurements of fibre bundles in the laboratory equipment and serves as comparison between those two techniques. The other method is to take an X-ray image at a certain temperature, return to room temperature, measure a second X-ray image and repeat this procedure stepwise to higher and higher temperatures. The intention is to clearly separate the permanent structural change from the thermal expansion. Constant temperature measurements (procedure 1) are therefore successive X-ray measurements at e.g. 1500 °C at a certain stress level for 1 h. Procedure 2 are individual X-ray measurements at constant stress and successive temperature levels like: 1000 °C - room temperature (RT) - 1200 °C - RT - 1400 °C - RT - et cetera. If not stated differently all measurements are performed with a 600 seconds multiread mode, comparing two X-ray diffraction images recorded for 300 seconds to avoid cosmic background.

For each measurement several steps are required:

4.2. The experiments

- Mounting of the fibre
- Moving the fibre into the X-ray beam
- Measurement of the electrical resistivity
- Room temperature X-ray measurement
- Temperature increase to test temperature
- Successive X-ray measurements
- End of the measurement

Mounting of the fibre The single carbon fibres are prepared as described in section 4.1. The gauge length of the fibre is measured with a caliper before the fibre is fixed to the sample holder, which is mounted into the vessel. The vessel is closed, the power line is connected and the vessel is evacuated. During the experiment the pressure is smaller than $1 \cdot 10^{-4}$ mbar.

Moving the fibre into the X-ray beam: It is quite difficult to find a fibre with a diameter in the range of microns in an X-ray with a similar size. To accelerate the procedure, the microspot beamline offers the use of a microscope, which can be positioned orthogonally to the X-ray beam. For the following procedure we made use of the glass window of the vessel, which was also oriented orthogonal to the beam. A FT500 fibre (strong X-ray signal) was mounted in the vessel and was moved until a strong X-ray signal was observed in the Detector. The fibre is then focussed in the microscope and thus the focus of the microscope corresponds to the position of the X-ray beam.

Then, for all other fibres the following technique was applied to move the fibre into the X-ray beam: The microscope was kept at its position and only the vessel was moved. If the respective fibre was visible in the focus of the microscope, it was also already very close to the X-ray beam and only a scan of $\pm 20\,\mu$m was required to position the fibre in the centre of the X-ray beam.

The whole procedure of moving fibres in and out and mounting new fibres requires the change of the positions of beam stop, detector and vessel. A script program was written, which performs the necessary steps in the right sequence.

4.2. The experiments

Measurement of the electrical resistivity: Low voltage is applied to the fibre. Current and voltage are measured and used to calculate the fibre diameter as described in 4.2.4. The values are typically about 2.5 V and 0.35 mA for the voltage and the current, respectively.

Room temperature X-ray measurement: For comparison with the data at high temperatures, a first X-ray diffraction image of the sample is taken at room temperature.

Temperature increase to test temperature: The voltage is successively increased and measured voltage and current are used to calculate the actual temperature. This is performed until the test temperature is reached. The temperature value is kept constant by adjusting the voltage.

Successive X-ray measurements: After reaching the test temperature X-ray diffraction images are recorded according to the measurement procedure chosen.

End of the measurement: After the maximum temperature for stepwise measurements, or after 1 h for constant creep measurements, the temperature is decreased until room temperature is reached again. A final X-ray diffraction image is made *post creep* at room temperature. The sample vessel is vented, the fibre is unmounted and preserved for further investigations. The fibre diameter and the mass of the dead weight are determined at the SEM and a high precision balance in Vienna, respectively. Values are listed in table 4.3.

4.2.6 Overview of the single fibre measurements

During a first "in-house" beam time the setup was tested and first preliminary experiments were performed. Single HTA5131 fibres were tested with in-situ X-ray diffraction, with the focus on the creep process. It turned out that the self mixed graphite adhesive sometimes failed to give an electrical contact sufficiently high to heat the fibres by direct electric current. Thus, only few fibres could be reliably measured. The second difficulty was the weak scattering signal, which led to difficulties in the evaluation and the background subtraction, as the signal from the fibre in comparison to the one from the *Kapton* window and the air was quite weak. Numerous tests were performed in the laboratory to improve the vessel, in particular the clamping procedure.

4.2. The experiments

For the second beam time, the following improvements enabled the in-situ measurements of single fibres: The beam intensity was increased by at least on order of magnitude by the microspot beamline with a new monochromator and a focussing system. The use of the silver glue as described in section 4.2.1 enabled direct heating with identical resistance from one experiment to another. The contact problems were solved by reconstructing the sample holder within the vessel, which was additionally positioned asymmetrically to measure the 002-reflection and the 10-band simultaneously.

The second beam time focussed on the comparison of the HTA5131 fibres, either as-received or heat treated at 1800 °C for two hours without load. A second focus was laid on the thermal expansion of the nanocrystallites within pitch-based carbon fibres, as the preliminary results from the first beam time already have shown that nearly no permanent structural change takes place in these fibres up to temperatures of about 2000 °C.

An overview on all single fibre experiments can be found in table 4.3. In

Nr. / #	Type	σ / MPa	T_{max} / °C	$\Delta_{T,max}$ / °C	D_m / μm	L / mm	Exp
S1	HTA	(260)	(1230)	15	(6.8)	15.0	step
S2	HTA	(260)	(1500)	30	(6.8)	15.2	step
S3	FT500	(170)	(1600)	50	(6.8)	16.1	therm
M1	HTA	160	1230	90	6.89	16.7	step
M3	K321	96	1650	20	9.20	14.0	therm
M4	K137	110	1760	15	8.38	15.9	therm
M5	HT1800	220	1850	20	6.53	15.5	step
M6	HT1800	230	1856	25	6.24	14.5	therm
M8	HTA	45	1470	50	6.67	14.5	step
M9	HTA	190	1350	80	6.56	14.1	const
M14	HTA	175	1200	140	6.34	14.8	const
M16	FT500	110	1470	25	8.48	15.2	therm
M20	FT500	(110)	(1630)	20	(8.50)	14.1	therm

Table 4.3: Relevant data of the single fibre measurements: Number of experiment, fibre type, applied stress σ, maximum temperature during the experiment T_{max} and its error $\Delta_{T,max}$, fibre diameter determined by SEM measurements D_m, gauge length of the fibre L and type of experiment performed. The numbers in brackets indicate fibres, which diameters were not measured in the SEM or fibres with graphite adhesive as described in the text. The error in the stress is in the order of 2%, for the fibre diameter it is smaller then 1%, and for the gauge length it is 0.1 mm. The different types of experiment are *step* and *const...*, for creep experiments with stepwise temperature change or constant temperature, respectively and *therm...*, for thermal expansion experiments.

4.2. The experiments

the table, the number of the experiment, the fibre type, the applied stress, the maximum test temperature together with its error, the diameter from the SEM, the gauge length and the experimental procedure is presented. The values in brackets indicate that either the diameter was not directly measured in the SEM or a graphite adhesive was used instead of the silver adhesive. As this affects also the determination of the applied stress level, this value is also written in brackets. The diameter of the samples S1, S2, S3 and M20 was taken as the mean value of these fibre types and the values of the applied stress and the temperature was calculated therefrom.

As the temperature calibration (section 4.2.4) of the carbon fibres was performed only for the silver adhesive, this procedure was adopted somehow for the samples S1, S2 and S3 with the graphite adhesive: Assuming that the resistance of the carbon fibres is identical and the only difference arises from the glue, the resistance of the graphite glue was calculated from the room temperature measurements R_{meas} of the resistance, $R_{glue} = R_{meas} - R_{th}$, where R_{th} are the corresponding values from the calibration of the Sigraflex-silver adhesive system. As the voltage U_{meas} and the current I_{meas} have been measured at each test temperature, the actual voltage on the fibre U_{fib} is calculated from $U_{fib} = U_{meas} - R_{glue} \cdot I_m$. This is based on the assumption that the resistance of the graphite adhesive is not changed during during the experiment, which is reasonable, as the large Sigraflex platelets remain nearly at room temperature. With U_{fib} and I_m the power and the corresponding actual temperature of the fibre are calculated using equation (4.2). Additionally the validity of these assumptions was checked by comparing the results from the measurements of the thermal expansion coefficient of the FT500 fibre for the sample S3 (with graphite glue and the above described evaluation) and M16 and M20 (with silver adhesive), which showed a very good correspondence.

The sample M8 was tested at a very low stress level, which is sufficiently low to ensure a straight loaded fibre, but on the other hand low enough to be well below the creep threshold. This sample serves as reference sample to the fibres loaded with different high stress levels to distinguish between creep and stress graphitisation and only thermal effects. The permanent changing stress graphitisation, differently to the reversible thermal elongation, was already indicated during the experiment by the electrical resistance, which decreased for loaded PAN-based fibres, where the voltage had to be permanently adjusted.

For the three measurements M3, M9 and M14 the planned temperature could not be reached, as for these experiments a leakage appeared, which was probably due to a problem with the turbomolecular pump of the vacuum system. Thus, oxidation damage inhibited longer measurement times and as a consequence the large number of temperature steps required to reach higher temperatures.

Small differences in the applied stress level are due to the different fibre diameters: The Sigraflex platelets were prepared to give the same stress value for the nominal fibre diameter, i.e. the mean fibre diameter of the bundle. As the size of fibres differs in a small range, this influences the stress, which was evaluated after the experiment by measuring the mass of the Sigraflex-adhesive-Sigraflex compound together with the precise diameter of the fibre in the SEM. With these values, the actual stress on the fibre during the experiment was recalculated.

4.3 Results

4.3.1 Structural change during in-situ creep

The following results are obtained from the evaluation of the two single fibre experiments S1 and S2. The interlayer spacing d_{002}, the azimuthal angle distribution of the pores $HWHM_p$ and the one of the graphene sheets $HWHM_g$ are determined in dependence on temperature. In figure 4.11 the change in the inter layer spacing with test temperature is depicted. The plot shows room temperature values and values obtained at high temperatures. All room temperature measurements are performed after cooling down the sample from a high temperature measurement. The test temperature is increased during the experiment. Both fibres clearly show the decreasing of the inter-layer spacing d_{002} with increasing test temperature. The change of d_{002}, and thus, the structural change is permanent because each measurement is performed at the same condition i.e. at room temperature, after heating to a certain temperature.

In figure 4.12 the HWHM evaluated from the azimuthal intensity distribution of the 002 peak is shown. This parameter describes the mean orientation of the graphene layers with respect to the fibre axis. The relatively large errors arise from the fit. It is assumed that the HWHM decreases with increasing test temperature, but as the decrease is within the error margin, no precise statement on the amount of decrease is possible. The two measurements at room temperature are performed before and after

4.3. Results

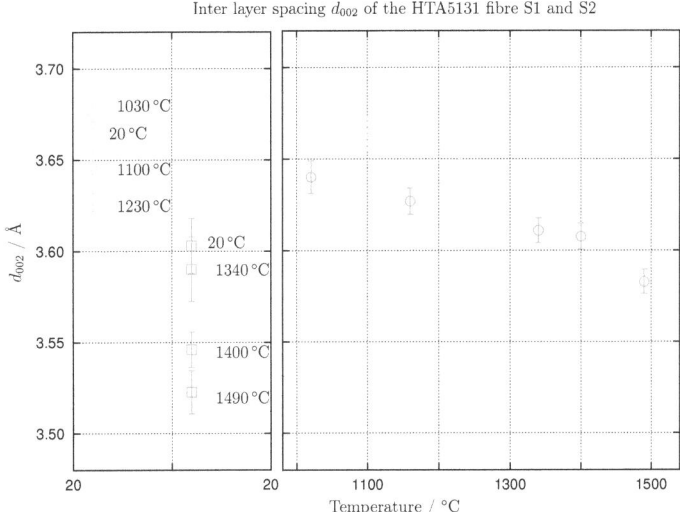

Figure 4.11: Inter-layer spacing d_{002} in dependence on temperature. The x-axis is plotted with intersections. The red squares and circles refer to sample S2, the green triangles to sample S1, respectively. The triangles pointing downwards and the squares indicate the results of measurements at room temperature, after being cooled down from the temperature, which is stated in the legend close to the symbol. Measurements at high temperatures are indicated with the triangles pointing upwards and the circles, respectively. d_{002} is decreasing with increasing test temperature.

the high temperature test and indicate a slight decrease in the HWHM.

Figure 4.13 shows the HWHM evaluated from the azimuthal intensity distribution of the SAXS signal. This parameter describes the mean orientation of the pores with respect to the fibre axis. Each of the room temperature values is measured after cooling down the sample from high temperatures as indicated in the figure at each point. The room temperature values and the values obtained by the high temperature measurements, respectively, decrease with increasing test temperature. The values at room temperature show that the structural change is permanent. The pores in the fibre exhibit a higher orientation with respect to the fibre axis with increasing test temperature.

4.3. Results

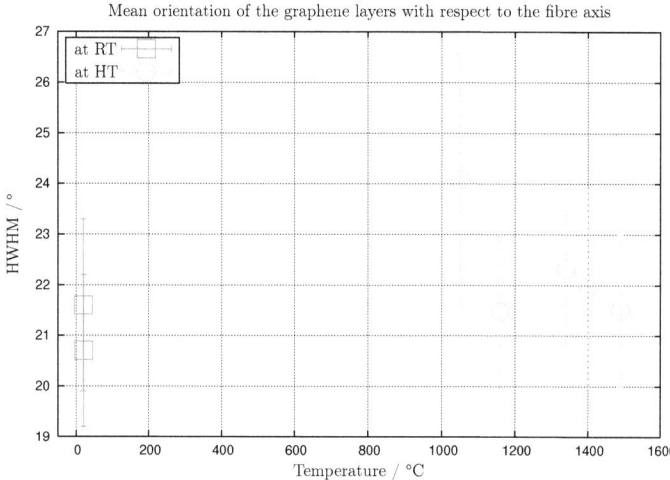

Figure 4.12: The HWHM of the 002 peak is shown for fibre sample S2. The red squares refer to the measurement at room temperatures (RT) before the first and after the final high temperature measurement. The green circles refer to the values measured at high temperatures.

4.3.2 Thermal expansion coefficient

The thermal expansion of the samples S3, M4 and M20 was measured and for the latter two samples the intensity of the X-ray signal was high enough to evaluate the thermal expansion of the crystallites with a size of some nanometers in all spatial directions. These directions are, with reference to the graphene crystallite, the out of plane direction c obtained from the 002-reflection and the in plane directions in fibre axis $a_\parallel$ and perpendicular to it $a_\perp$, obtained from the 10-band parallel to the fibre axis (and the applied stress) and perpendicular to it, respectively. The directions are depicted in figure 4.14 with respect to the orientation of the fibre.

In the measurements the change of the lattice parameters c, $a_\parallel$ and $a_\perp$ is obtained. For the evaluation of the thermal expansion coefficients a linear fit is applied to each data set and the thermal expansion coefficients are calculated using equation (2.1).

Thermal expansion in out of plane direction

In figure 4.15 the inter-layer distance d_{002} of the graphene planes of the samples S3, M20 (both FT500 fibres) and M4 (a K137 fibre) is shown in

106

4.3. Results

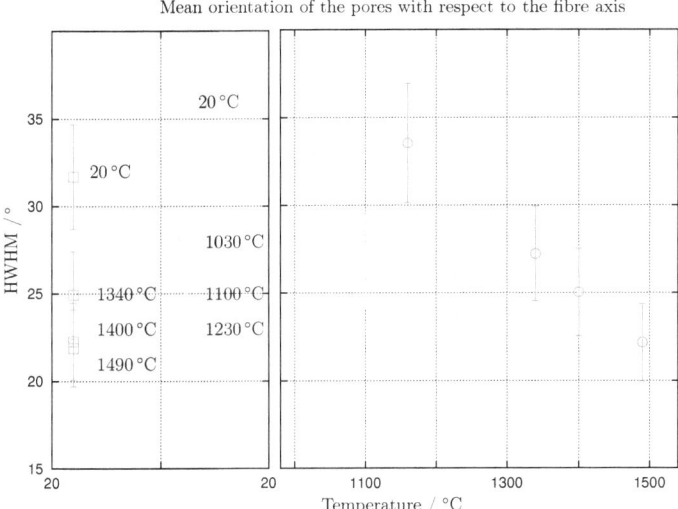

Figure 4.13: The HWHM from the SAXS-signal is shown for fibre sample S1 (green triangles) and sample S2 (red squares and circles), respectively. The x-axis separates the room temperature (RT) values (down pointing triangles and squares) and values obtained during high temperature creep (HT). The room temperature measurements have been performed after cooling down the sample from the indicated temperature.

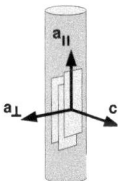

Figure 4.14: Scheme of the crystallographic directions of the the graphene layers within the carbon fibre, for which the thermal expansion has been evaluated.

dependence on temperature. After each test temperature, the fibre was cooled down and an X-ray pattern was taken at room temperature. No structural change was observed, all room temperature values are identical with respect to the error margin. A linear fit was applied to each data set and the thermal expansion coefficient was calculated using equation (2.1) with an initial length of $l_{S3,M4} = 3.43$ Å and $l_{M20} = 3.42$ Å, respectively. For S3 the thermal expansion is linear in the investigated temperature

4.3. Results

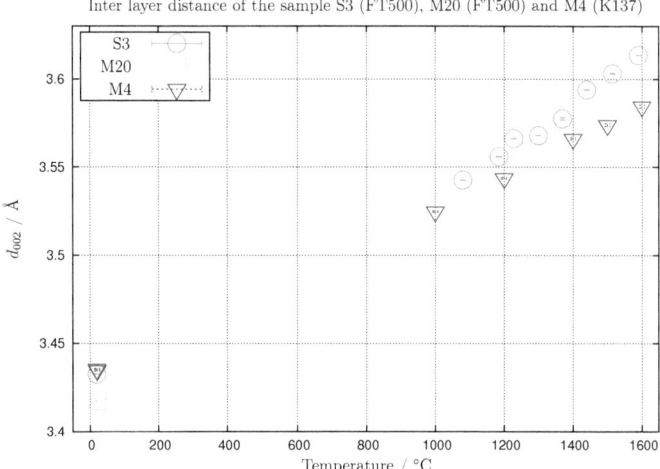

Figure 4.15: Inter-layer distance d_{002} in dependence on temperature. Measurements at room temperature performed before and after the high temperature test showed identical values.

range from 1000 °C to 1600 °C, i.e. the thermal expansion coefficient has a constant value of:

$$\alpha_{c,FT500,S3} = (4.1 \pm 0.2) \cdot 10^{-5} \, K^{-1}$$

Sample M20 was evaluated in two different temperature ranges, for temperatures lower than 800 °C and for temperatures higher than 800 °C, respectively. In the higher temperature range the thermal expansion coefficient was found to be:

$$\alpha_{c,FT500,M20,high} = (3.01 \pm 0.03) \cdot 10^{-5} \, K^{-1}$$

and in the lower temperature range:

$$\alpha_{c,FT500,M20,low} = (1.6 \pm 0.2) \cdot 10^{-5} \, K^{-1}$$

The thermal expansion coefficient for the FT500 fibre is increasing between room temperature and 800 °C. The thermal expansion coefficient for the K137 fibre is linear for temperatures higher than room temperature, because all data points, also the value determined at room temperature can be adjusted by one linear fit. The coefficient evaluated is given by:

4.3. Results

$$\alpha_{c,M4} = (2.75 \pm 0.03) \cdot 10^{-5} K^{-5}$$

All values are of comparable size and the values of the two different FT500 samples are similar. The linear expansion coefficient in out of plane direction of the graphene crystallites was found to increase with increasing temperature for the FT500 fibres while the value of the K137 fibre was found to be linear for temperatures higher than room temperature. The value of the K137 fibre lies between the low and the high temperature value of the FT500 fibre and is comparable to the mean value of both.

Thermal expansion in plane parallel to the fibre axis

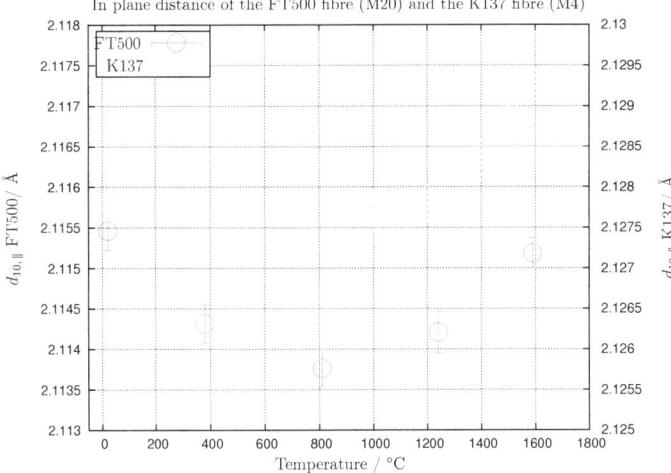

Figure 4.16: In plane distance of the carbon atoms in $a_{\parallel}$-direction (parallel to the fibre axis). The red circles refer to the left scale, while the green squares refer to the right scale. Both scales have the same range, they differ only in the offset.

High precision is required for the evaluation of the in plane thermal expansion. The obtained values are very reliable, because the error-bar of the fit applied during data evaluation is small and the peak of the scattering curve I(q) versus q is very well defined.

The change in the interatomic distances in plane are of the dimension of 0.002 Å, (which is of the order of 0.09 % of the numerical values), whereas

109

4.3. Results

the precision of the fit is of the dimension 0.0005 Å.

In figure 4.16 the in plane lattice distance $d_{10,\parallel}$ of the carbon atoms in the graphene crystallites is shown. Although no measurements for the K137 fibre between room temperature and 1000 °C are considered, it is obvious that both fibres show different behaviour.
The room temperature value of the FT500 fibre is initially 2.1155 Å. The values for higher temperatures decrease for temperatures below 800 °C and increase again for higher temperatures.

A linear fit was applied to values at temperatures higher than 800 °C and the thermal expansion coefficient was determined using equation (2.1) with $l_{\parallel,\text{FT500}} = 2.1155\,\text{Å}$:[3]

$$\alpha_{\parallel,\text{M20, high}} = (0.9 \pm 0.2) \cdot 10^{-6}\, K^{-1}.$$

The coefficient determined for temperatures lower than 800 °C is:[4]

$$\alpha_{\parallel,\text{M20, low}} = (-1.0 \pm 0.3) \cdot 10^{-6}\, K^{-1}.$$

The K137 fibre shows different behaviour. The $d_{10,\parallel}$ values appear to be constant between room temperature and 1000 °C and increase then with increasing temperature, exhibiting linear behaviour between 1000 °C and 1700 °C. The thermal expansion coefficient for the K137 fibre was calculated from a linear fit to values at temperatures higher than 1450 °C. Then equation (2.1) was used with $l_{\parallel,\text{K137}} = 2.1158\,\text{Å}$:

$$\alpha_{\parallel,\text{M4}} = (1.8 \pm 0.1) \cdot 10^{-6}\, K^{-1}$$

The thermal expansion coefficient in plane of the graphene crystallite is at least ten times smaller than the one found for the out of plane direction. No statement can be made about the region below 1000 °C but it is supposed that the values are constant as well. For both fibre types, the room temperature value is nearly unchanged after the high temperature measurements.

Thermal expansion in plane perpendicular to the fibre axis

In figure 4.17 the distance of the carbon atoms, evaluated from the 10-band in $a_\perp$-direction (perpendicular to the fibre axis) is shown. All values exhibit

---

[3]If the calculation is performed by only taking the two highest temperature steps into account, a 1.6 times higher result is obtained.

[4]If the calculation is performed by only taking the values at 400 °C and 800 °C into account, a 1.7 times smaller result is obtained.

4.3. Results

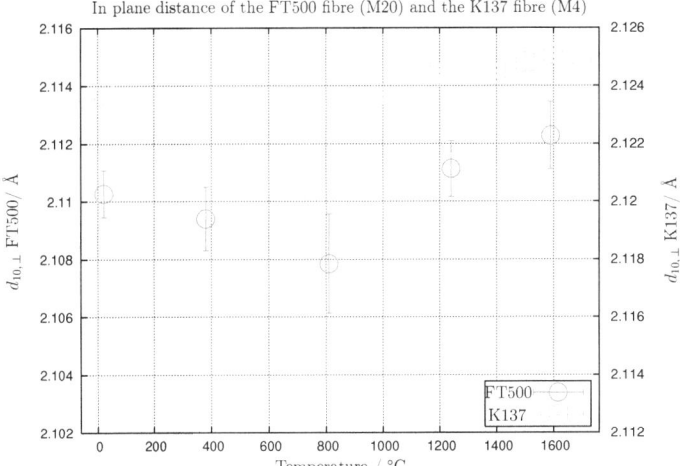

Figure 4.17: In plane distance of the carbon atoms in a$_\perp$-direction (perpendicular to the fibre axis). The red circles refer to the left scale, while the green squares refer to the right scale.

a significantly higher error then the corresponding values in a$_\parallel$-direction, because the scattering intensity is much smaller in this case, which leads to an increase of the fit error.

The in-plane atomic distance $d_{10,\perp}$ for the FT500 fibre decreases with increasing testing temperature for temperatures below 800 °C and increases again for higher temperatures. A linear fit was applied to data at temperatures higher than 800 °C and the thermal expansion coefficient was calculated using equation (2.1) with an initial length of $l_{\perp,\text{M20}} = 2.11$ Å:

$$\alpha_{\perp,\text{M20, high}} = (2.7 \pm 0.6) \cdot 10^{-6}\, K^{-1}.$$

From the elongation at 400 °C and 800 °C a thermal expansion coefficient for low temperatures was evaluated using the same reference length:

$$\alpha_{\perp,\text{M20, low}} = (-1.7 \pm 0.3) \cdot 10^{-6}\, K^{-1}.$$

The $d_{10,\perp}$ for the K137 fibre appears to be constant between room temperature and 1000 °C, whereas it increases for higher test temperatures. The value for 1700 °C is slightly smaller than the value at 1600 °C, but it was difficult to evaluate and is therefore questionable - a higher error has to be

4.3. Results

taken into account. The thermal expansion coefficient was evaluated applying a linear fit to the three values between 1400 °C and 1600 °C, using equation (2.1) with $l_{\perp,\text{K}137} = 2.1237\,\text{Å}$:

$$\alpha_{\perp,\text{M}4} = (1.7 \pm 0.3) \cdot 10^{-6}\,K^{-1}.$$

The value is similar to the one found in $a_\parallel$-direction.

A similar behaviour is observed for the in-plane thermal expansion parallel, as well as perpendicular to the fibre axis, which leads to the conclusion that the properties within the plane are identical and there is no preferred diction.

Table 4.4 gives an overview on all the calculated thermal expansion coefficients.

Fibre	Dir	α $\cdot 10^{-6}\,K^{-1}$	$\Delta\alpha$ $\cdot 10^{-6}\,K^{-1}$	Comment
HTA5131	c	35	2	bundle
FT500	c	41	2	single fibre
FT500	c	30.1	0.3	single fibre HT
FT500	c	16	2	single fibre LT
K137	c	27.5	0.3	single fibre
FT500	$a_\parallel$	−1.0	0.3	single fibre LT
FT500	$a_\parallel$	0.9	0.2	single fibre HT
K137	$a_\parallel$	1.8	0.1	single fibre
FT500	$a_\perp$	−1.7	0.3	single fibre LT
FT500	$a_\perp$	2.7	0.6	single fibre HT
K137	$a_\perp$	1.7	0.3	single fibre

Table 4.4: Thermal expansion coefficients for the fibres FT500 and K137 for the direction (Dir) c (out of plane), $a_\parallel$ and $a_\perp$ (both in plane) of the hexagonal lattice. HT and LT indicate evaluations using values higher respectively lower than 800 °C.

The FT500 fibre has a non linear thermal expansion coefficient in all three crystallographic directions. The thermal expansion coefficient is positive in all directions for temperatures higher than 1000 °C, but is negative for the in-plane directions for temperatures between room temperature and 800 °C. The K137 fibre shows totally different behaviour. It exhibits linear thermal expansion in all three crystallographic directions for temperatures higher than 1000 °C.

Chapter 5
Post-Creep measurements

This section summarizes the results, which were published in [67] within the frame of this thesis.

5.1 Experimental procedure

Post-creep measurements with X-rays were used to investigate residual changes of carbon fibres due to creep. The influence of load and temperature on the structure was determined, but it is not possible to get information on the time dependence of the underlying process. The main scientific goal was to compare (non-stabilised) as-received fibres to stabilised fibres, which were heat-treated at 2100 °C for two hours without load. The fibres were subjected to different stress levels of 173, 217 and 260 MPa at temperatures between 1500 °C and 1800 °C. The tests were performed in a hydraulic testing machine at constant load, the elongation of the fibre-bundles was determined from the cross-head displacement. For the evaluation of the data, the simplified Dorn-equation (2.2) was used in the steady state creep region. After the creep experiments the fibre-bundles were investigated by X-rays. Therefore the bundles were taken out of the grips, fixed on a appropriate sample holder and SAXS and WAXD measurements were performed.

5.2 Results

In this section the most important post creep results from [67] are presented. Creep experiments (high temperature and high load applied for about 90 min to a fibre bundle of the type HTA5131) have been performed with as received fibres (non-stabilised) and fibres, which were heat treated at 1800 °C for two hours (stabilized), respectively. The same measurements procedure is applied to all samples and the structure of the samples was investigated using X-ray diffraction after the high temperature experiments.

The main results are presented in figures 5.1 and 5.2 and show the difference in the mean orientation of the graphene layers $HWHM$ evaluated from the azimuthal distribution of the 002-reflection and the radius of gyration R_g, respectively, both in dependence on the temperature applied during creep.

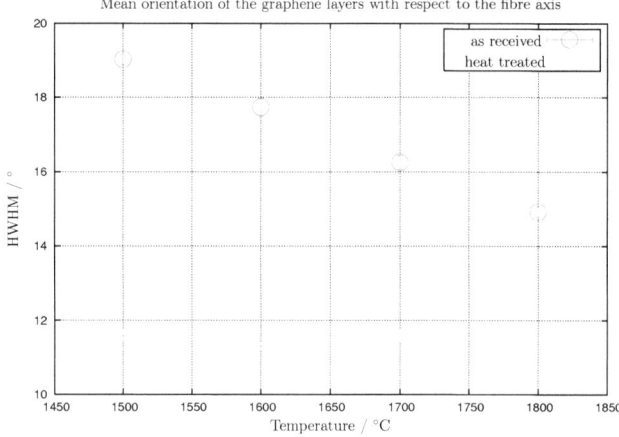

Figure 5.1: HWHM of the graphene planes of the non-stabilized fibres (red circles) and the stabilized fibres (green squares) at a stress level of about 217 MPa, in dependence on temperature.

The post creep values of the $HWHM$ decrease with increasing test temperature for the non-stabilized fibres, whereas the ones of the stabilized fibres are not affected by the heat treatment under load and remain at a much lower level (about 11.5°) for all temperatures, as shown in figure 5.1. A similar effect can be found for the radius of gyration (figure 5.2). The values of the

5.2. Results

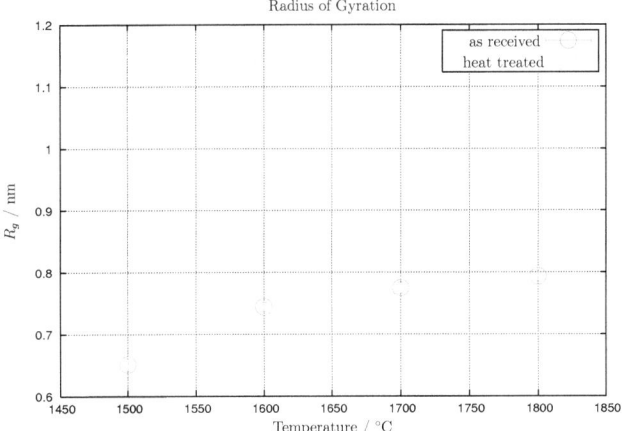

Figure 5.2: R_g of the pore cross-section of the non-stabilized fibres (red circles) and the stabilized fibres (green squares) at a stress level of about 217 MPa in dependence on temperature.

non-stabilised fibres increase with increasing testing temperature, whereas the ones of the stabilized fibres remain constant (about 1.1 nm) at a much higher level.

It is clear that the permanent structural change is temperature dependent. The higher the temperature during the experiment, the higher the graphene plane orientation with respect to the fibre axis and the higher the radius of gyration of the pore cross section (perpendicular to the fibre axis).

Chapter 6

Discussion

In-situ creep measurements have been performed with single carbon fibres and with carbon fibre bundles, respectively. Carbon fibre bundles turned out to be better suited for the evaluation of mechanical parameters such as the activation energy, the activation volume and the creep parameter, because one single test measures at the same time the statistical average of thousands of single fibres. The development of structural parameters such as the mean orientation of the graphene crystallites with respect to the fibre axis, the crystallite size, the pore radius, or the inter-layer spacing could be followed with the X-ray transparent setup for tensile tests at high temperatures and high stresses, developed within the frame of this thesis.

On the other hand, single fibre in-situ experiments allow a much more accurate measurement of the development of structural parameters in dependence on time and temperature. The thermal expansion of crystallites consisting of a small number of graphene sheets, with a size of only some nanometers, can only be measure with single fibres in the microbeam of a synchrotron radiation source. Thus, considerable differences in the structure - mechanics relation of fibres originating from different precursors (PAN and MPP) are revealed.

6.1 Thermal expansion of nanocrystallites

There are only few publications about the thermal expansion of carbon fibres [69], measuring the elongation of the fibre or its diameter in dependence on temperature, and thus the macroscopic thermal expansion. These values are as an average of all crystallites, not directly comparable [20] to values obtained by measurements on pyrolytic graphite [35] or thin carbon films [79]. The fibre structure, e.g. the orientation of the graphene crystal-

6.1. Thermal expansion of nanocrystallites

lites with respect to the fibre axis, would have to be taken into account. However, there is a considerable lack in data for these nancrystallites - there have been, to the knowledge of the author, no explicit measurements of the thermal expansion of the graphene crystallite in a carbon fibre, and thus the microscopic thermal expansion.

A macroscopic value for the longitudinal thermal expansion (in fibre axis) was published recently [71] and is of the same dimension as the value found in this work. Macroscopic values for the radial thermal expansion (perpendicular to the fibre axis) have been measured by TEM [69]. All values are listed in table 6.1.

In this work, the thermal expansion coefficients of carbon nanocrystallites were measured by X-ray diffraction of single carbon fibres. The values obtained can therefore be compared to values found for bulk graphite [2] or pyrolytic graphite [22]. The large anisotropy between the thermal expansion out of plane and the thermal expansion in plane of the lattice of the graphite crystallite is also documented in the literature [20]. Values measured for the microscopic thermal expansion are listed in table 6.1.

	Macroscopic				Microscopic		
Dir	α_1 $\cdot 10^{-6} K^{-1}$	α_{lit} $\cdot 10^{-6} K^{-1}$	cit	Dir	α_2 $\cdot 10^{-6} K^{-1}$	α_{lit} $\cdot 10^{-6} K^{-1}$	cit
long	1.83	2.0	[71]	a_{HT}	0.9	0.95	[2]
trans	–	30.0	[69]	a_{LT}	-1.0	-1.5	[2]
				c	30.1	28.09	[2]

Table 6.1: Thermal expansion coefficients comparing macroscopic and microscopic values found in this work (α_1 and α_2) to values from the literature α_{lit}. α_1 was measured for a PAN based HTA5131 fibre and the values α_2 for the sample M20, a pitch based FT500 fibre. Dir denotes the direction, which in the macroscopic case is indicated (*long*) and (*trans*) for the longitudinal and transversal direction, respectively, referring to the fibre axis. In the microscopic case a_{HT} and a_{LT} indicate the in plane lattice distance of the carbon atoms in the graphene crystallites at high and low temperatures, respectively, whereas c indicates the out of plane direction. The literature value for the longitudinal thermal expansion was measured at about 1000 °C. The literature value for a_{LT} was determined at 150 °C, while the literature values for a_{HT} and c were determined in a temperature range from 1000 °C to 1800 °C. The values measured in this thesis correspond well to the observations in the literature for bulk graphite.

In figure 6.1, the values of the macroscopic and microscopic thermal expansions are compared for PAN and MPP fibres, respectively. The in plane microscopic thermal expansion was only determined for the MPP fibres FT500 and K137 is presented in figure 6.2.

6.1. Thermal expansion of nanocrystallites

For perfectly aligned crystallites with respect to the fibre axis, the macro-

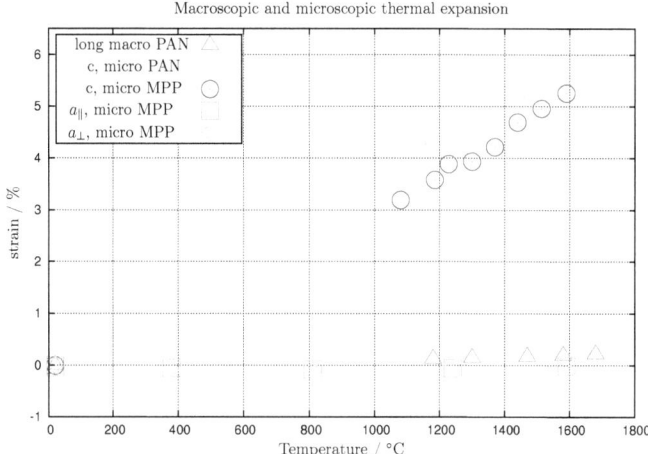

Figure 6.1: Macroscopic and microscopic thermal expansion for PAN fibres (triangles) and MPP fibres (circles and squares), respectively. The strain is calculated by dividing the lattice constant (c, $a_\parallel$ or $a_\perp$) evaluated from the X-ray diffraction pattern at a given temperature by an initial value, obtained from a measurement at room temperature: $l_{temp}/l_{initial}$. The microscopic thermal expansion of the PAN fibre and the MPP fibre are of the same dimension in c direction. Values for $a_\parallel$ and $a_\perp$ are small in comparison to out of plane values and are shown enlarged in figure 6.2.

scopic longitudinal thermal expansion of the carbon fibre would be related to the in plane thermal expansion of the graphene crystallite at high temperatures, while the macroscopic transversal thermal expansion of the fibres would be related to the out of plane thermal expansion of a graphene crystallite. The macroscopic and microscopic values can be related [20] by:

$$\alpha_{long} = \alpha_c \cdot sin\phi + \alpha_a \cdot cos\phi \qquad (6.1)$$

with ϕ being the mean angle of misalignment which can be determined e.g. by X-ray diffraction. ϕ is approximately 20° for as-received PAN fibres and considerably smaller for heat treated PAN fibres or most of the MPP fibres, e.g. about 6.7° for the FT500 fibre [52]. The macroscopic longitudinal thermal expansion for temperatures higher than 1000 °C was calculated applying equation (6.1) to the microscopic thermal expansion coefficients listed in table 6.1. The result of 4.6 K^{-1} is of the same dimension as the macroscopic values listed in the table, which shows that the macroscopic

6.2. Creep of carbon fibres

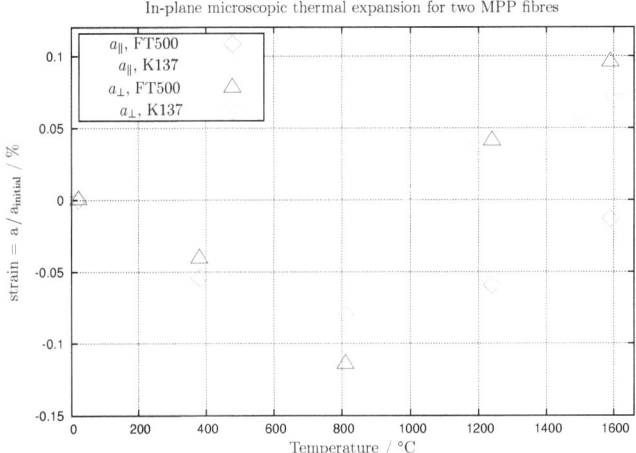

Figure 6.2: Microscopic thermal expansion for the two MPP fibres FT500 and K137. While the FT500 fibre shows negative thermal expansion for temperatures below 800 °C, this is not observed for the K137 fibre.

and microscopic values obtained in this work correspond to each other.

In the literature [20] results on the macroscopic thermal expansion of ex-PAN as well as ex-pitch fibres show shrinking between room temperature and 1000 °C up to a strain of 0.04 %. In PAN fibres this effect is less pronounced and only observed for temperatures below 400 °C [71]. In our experiments, we did not observe this behaviour for the fibre bundles HTA5131 investigated, see figure 6.3. (The resolution of the strain measurement was about 0.003 %.) The fibres showed macroscopic elongation in direction of the fibre axis for all temperatures due to a successive and permanent change of the fibre structure (increasing orientation of the crystallites with respect to the fibre axis).

6.2 Creep of carbon fibres

Non stabilized PAN based carbon fibres of the type HTA5131 were investigated by in-situ X-ray diffraction and show permanent structural change due to a creep like process. The pore radius perpendicular to the fibre axis R_g as well as the crystallite size L_c increase with increasing time in the creep experiment. A higher orientation of the pores and crystallite ribbons

6.2. Creep of carbon fibres

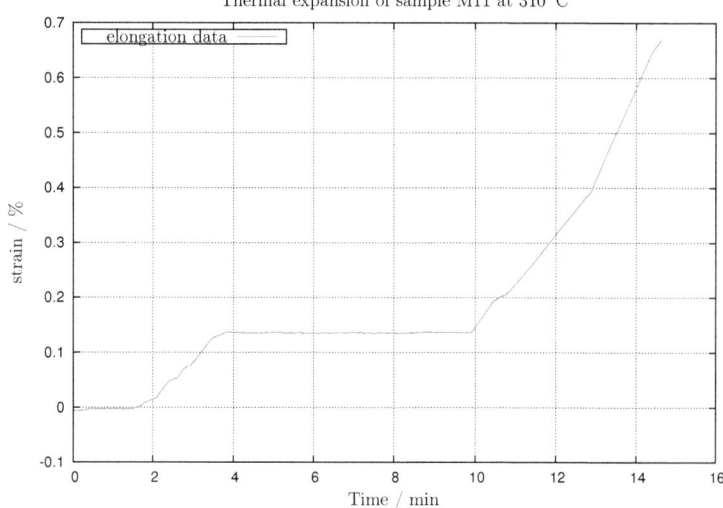

Figure 6.3: Macroscopic thermal expansion of a HTA5131 fibre (sample M11) - original data. The fibre bundle was heated from room temperature to 310 °C and the temperature was kept constant for 6 minutes. The fibres show positive thermal expansion, no shrinking of the fibres is observed.

with respect to the fibre axis observed: Figure 6.4 shows selected intensity distributions of the 002 peak as a function of the azimuthal angle χ as an example.

It was found, that the main structural change observed due to the creep process is achieved within only five to ten minutes of heat treatment. All parameters drastically change in the time period of the primary creep, and change at a much smaller rate in the period of secondary creep.

The structural change depends on temperature and time of heat treatment. The fibre bundle measured at 1500 °C and 20 MPa is only weakly affected by stress and temperature but the trend, however, is similar: The orientation distribution of the graphene sheets with respect to the fibre axis decreases with increasing time of the experiment (figure 3.25). This indicates that the fibres which were heat treated at high temperatures but without load, feature a structural change at a much lower rate. E.g. the HT1800 fibres, investigated in [67], were heat treated at 1800 °C without any load for two hours and, however, show highly oriented graphene layers and pores as well as a highly increased value of the radius of gyration.

6.2. Creep of carbon fibres

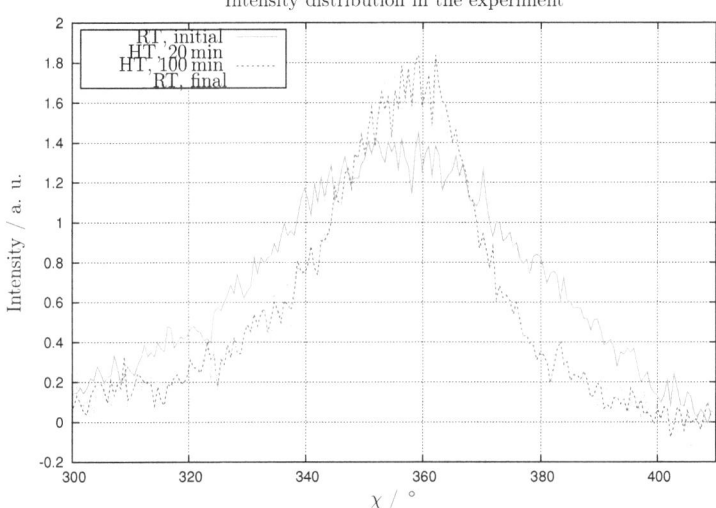

Figure 6.4: Intensity distribution in dependence on the azimuthal angle χ. $RT, initial$...measurement at room temperature before the experiment, $HT, 20\,min$ and $HT, 100\,min$...measurements after 20 minutes and 100 minutes, respectively, $RT, final$...room temperature measurement after cooling down the sample at the end of the experiment. Loading and temperature conditions were 260 MPa and 1500 °C (HT), respectively.

The inter-layer spacing of the graphene crystallites decreases with increasing time at high temperatures. The change of the fibre structure can be seen in the X-ray diffraction images: Figure 6.5 shows the intensity distribution of the 002 peak as a function of the inter-layer spacing d_{002} as recorded in the experiment.

Immediately after heating, the peak of the 002 reflection is considerably smaller and its height increases. Its position shifts towards smaller q-values (higher d-values) due to thermal expansion. Only a small further shift of the peak with time is observed. After cooling down to RT, a final X-ray measurement shows that the change of d_{002} is permanent. The radial intensity distribution of the 002-reflection has shifted to lower d_{002}.

In figure 6.6 the inter-layer spacing d_{002} is presented for a FT500 fibre and a HTA5131 fibre, in dependence on the temperature. Heating the MPP fibre results in a linear increase of the inter-layer spacing, because the

6.2. Creep of carbon fibres

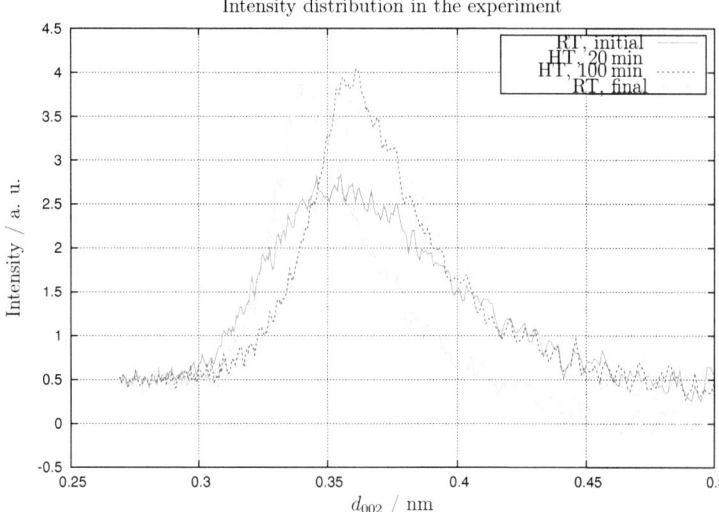

Figure 6.5: Intensity distribution in radial direction as a function of the inter-layer distance d_{002}. The curve measured at room temperature (RT) before the experiment is indicated *initial*, while the curve measured at RT after cooling down the sample after the experiment is indicated *final*. 1510 °C (HT) and 210 MPa were applied to the fibres.

crystallite size and the orientation of the crystallites is not changed by the heat treatment and thus only thermal expansion is observed. The inter-layer spacing of the PAN fibre first increases at temperatures up to 800 °C due to thermal expansion, but decreases for higher temperatures. Probably due to the coalescence of crystallites and rearrangement of graphene sheets within the fibre, d_{002} and thus the corresponding strain decreases. Measurements at room temperature showed that the heat treatment changed the inter-layer spacing of the PAN fibre permanently, which is concluded from the in-situ experiments at the synchrotron radiation source, shown in figure figure 4.11.

Also the mean orientation of the graphene crystallites with respect to the fibre axis, the crystallite size and the pore radius of the PAN based carbon fibres change permanently due to the creep process, which can be achieved by heat treating the fibre without load, but is accelerated by additionally applying load.

MPP fibres and PAN fibres heat treated at high temperatures are creep resistant. MPP fibres are produced from a different precursor material and

6.3. Structural model

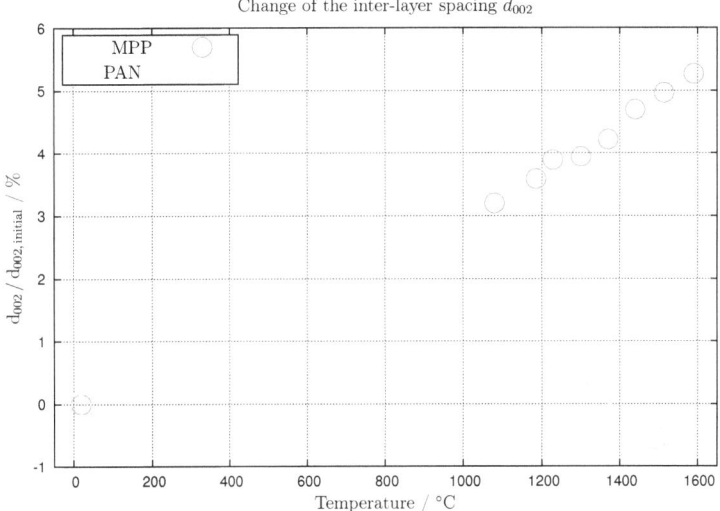

Figure 6.6: Relative change of inter-layer spacing for the MPP fibre FT500 and the PAN fibre HTA5131. d_{002} and thus the strain of the FT500 fibre shows a linear increase with increasing temperature, while the strain of the HTA5131 fibre decrease for temperatures higher than 800 °C, due to structural change in the PAN fibre.

the fibres consist of graphitic crystallites that are highly oriented. The crystallites of a heat treated PAN fibre tend to be oriented parallel to the fibre axis and the inter-layer distance of the graphene crystallites approximates the value of bulk graphite. Further structural change is only possible at higher heat treatment temperatures and thus, the fibre is creep resistant at low temperatures.

The measurements presented in this thesis show that heat treatment is a suitable method to achieve PAN fibres that are creep resistant up to high temperatures. Fibres treated at 1800 °C would not show any creep up to this temperature. Further it is shown that due to a stress induced processes only a short heat treatment time is sufficient, while the main change occurs in the primary creep phase.

6.3 Structural model

Different mechanisms of the creep process in carbon fibres can be distinguished. The creep process at temperatures higher than 2000 °C (high

6.3. Structural model

temperature creep) is based on vacancy formation and motion in the graphene lattice [76], or sliding of crystallites caused by bulk diffusion of individual interstitial carbon atoms [45]. The subject of the present work is the creep process at temperatures below 2000 °C (low temperature creep), which is characterized by a rearrangement of the basic structural units induced by stress and temperature.

For low temperature creep the value found for the microscopic activation energy $H = (2.0 \pm 0.3)$ eV is significantly lower than values for high temperature creep, e.g. for the fibre E130, an activation energy of $Q = (11.2 \pm 0.9)$ eV was observed [76]. Theoretical calculations proposed a similar value, i.e. $Q = 11.4,$ eV for diffusion of carbon atoms [41], [19].

The small activation energy is an indication of the not yet finished structural development in the non-stabilised fibres even at relatively low temperatures (less than 2000 °C). The observed value of the creep exponent $n = 2.0 \pm 0.3$ further suggests [15] that a reorganisation process takes place, which is probably grain boundary gliding of the crystallites, driven by an optimisation of the structure with respect to minimizing elastic energy.

The activation volume $\nu = (2.8 \pm 0.1) \cdot 10^{-28}$ m^3 is about 80 times smaller than the coherent scattering region (presented in figure 6.7), but its numerical value indicates a process, that includes a grain boundary between two crystallites.

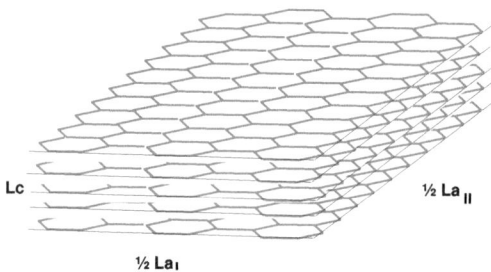

Figure 6.7: Dimension of the coherently scattering crystallite for the HTA5131 fibre: $L_{a,\parallel} \sim 4.7$ nm, $L_{c,\perp} \sim 3.4$ nm and $L_c \sim 1.6$ nm. The values were calculated using equation (2.12).

The activation volume with respect to the crystallites is presented in figure 6.8 illustrating a cube like and a flat cubic volume as examples: A crystallite with a small stacking height L_c glides from a neighbour crystallite, a large area contributing in the process, or two crystallites with large L_c but a small overlap region glide.

6.3. Structural model

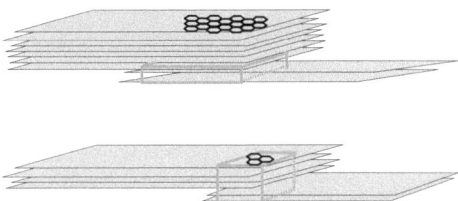

Figure 6.8: Two possible arrangements to demonstrate the relation activation volume - coherent scattering crystallite. The activation volume is indicated by the red cube, the corresponding in-plane area of the graphene sheets is shown by the hexagons.

For the low temperature the results (table 6.2) further allow to distinguish between a mainly stress dependent process that takes place for stress and temperature values higher than a certain creep threshold and a process only depending on temperature. Stress induced sliding of graphene sheets

	depending on		
Parameter	$\sigma > \sigma_{CT}$	T	$time$
$HWHM$	✓	✓	✓
R_g	✓	✓	✓
L_c		✓	✓
d_{002}			✓

Table 6.2: Parameters characterising the structure of a carbon fibre investigated in this work are: The half width at half maximum of the orientation distribution of the graphene crystallites with respect to the fibre axis $HWHM$, the radius of gyration R_g, the stacking height of the crystallites L_c and the inter-layer spacing d_{002}. The change of the parameter in the creep experiment was found to depend on whether or not the stress applied to the fibre σ was above a certain level σ_{CT}, on the temperature T and on the duration $time$. Individual parameters were increasing or decreasing with increasing stress, temperature and time as explained in the text.

as presented in figure 6.9 is suggested. Both models are possible, but the model presented on the left hand side better describes the higher orientation of the graphene sheets with respect to the fibre axis, which is observed in the experiments. Although the process is mainly depending on the stress, the temperature has to be high enough to allow the graphene sheet to move from one position with respect to the neighboring sheet to another energetically favourable position, which is only possible if an energy barrier due to local bonding is overcome. One possible scheme for pore growth caused by grain boundary sliding is shown in figure 6.10.

6.3. Structural model

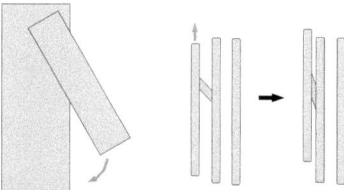

Figure 6.9: The stress applied to the fibres leads to shear forces, resulting in sliding of graphene crystallites or graphene sheets with respect to each other. On the left hand side the graphene sheet in the front slides across the one in the back by rotation. This increases the orientation with respect to the fibre axis. On the right hand side the crystallites slide along each other, which is increasing the pore volume of neighboring pores, but does not increase the crystallite orientation significantly.

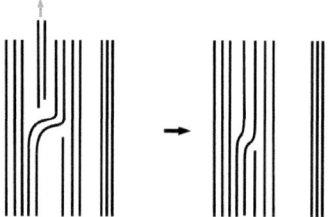

Figure 6.10: Scheme for grain boundary sliding. As the small crystallite is removed, the distorted crystallite relaxes. The pore size, an inclusion between two crystallites, is increased.

For both parameters, L_c and d_{002}, the degree and the rate of change in the creep experiment do not depend on the stress level, but L_c depends on the temperature and d_{002} depends on the time of high temperature treatment only. This indicates a thermally activated process in the fibre that does not depend on macroscopically applied stress. It is suggested, that neighbouring crystallites in an untreated PAN fibre are in an energetically unfavourable arrangement, stabilized by the crystallite network. Heating the fibre results in thermal elongation and in internal stresses, leading to a slow rearrangement.

Fibres heat treated without load at high temperatures show a highly oriented fibre structure and are found to be creep resistant for temperatures below the heat treatment temperature. From this observation, one concludes that after high temperature treatment without load, the structure of the fibre relaxes to an energetically favourable arrangement.

Untreated samples cooled to room temperature were heated, and the same temperature and stress was applied again. The fibres were creep resistant, although they showed usual secondary creep before cooling. It is suggested that the fibres heat treated under stress continue to rearrange crystallites until fracture, as the microstructure can not settle into a stable position, whereas it can settle into a stable arrangement if relaxed from tension and temperature.

PAN fibres of the type investigated contain about 4 weight percent of nitrogen and a small amount of hydrogen. These elements are split off during the high temperature treatment, which results in open bonds. This could also play an important role for the rearrangement process, as a consequence of thermal treatment of the fibres.

Chapter 7
Conclusion

Two testing devices have been designed and constructed, which are suited for in-situ X-ray diffraction together with creep experiments on carbon fibre bundles and single carbon fibres, respectively. The devices work in vacuum (pressure smaller than 10^{-4} mbar) and have been successfully used in the laboratory X-ray equipment (for fibre bundles) and in the synchrotron radiation source BESSY (for single fibres). Different types of carbon fibres have been investigated, PAN-fibres (HTA5131 as received, and stabilized by heat treatment) and MPP-fibres (K137, K321, FT500). Experiments showed that there exist two different creep processes for low temperatures, as investigated in this thesis and for high temperatures e.g. presented by Kogure [45], respectively. At temperatures below 2000 °C stress induced sliding of graphene crystallites and thermally activated rearrangement of the microstructure in the fibre are the leading mechanisms for the creep process.

The structure of HTA5131 carbon fibres is permanently changed by heat treatment and simultaneously applied load. As a consequence, the fibres become creep resistant for lower temperatures than the test temperature of the previous creep experiment. Except for the inter-layer distance d_{002}, all parameters related to the fibre structure show a strong response during primary creep, followed by slower change during the steady-state secondary creep.

The mean orientation of the graphene layers with respect to the fibre axis is found to increase with time and with increasing temperature and increasing stress. The inter-layer spacing within the graphene crystallite d_{002} is decreasing slightly with increasing time during the experiment, but at the same rate for primary and secondary creep. No correlation to the test temperature or the stress level is found. The cross section of the pores is determined by the radius of gyration R_g, which increases with time in the course of the experiment. It is also depending on the test temperature and the stress level. The crystallite size perpendicular to the fibre axis L_c is increasing during the experiment, showing correlation to the test temperature but not to the stress level.

The thermal expansion was investigated for the FT500 fibre and the K137 fibre in the c-direction out of plane and the the a-direction in plane parallel and perpendicular to the fibre axis, with the notation referring to the graphene crystallites.

The tensile tests of carbon fibres at temperatures up to 1800 degrees with simultaneously investigating the structural development by in-situ X-ray diffraction was experimentally challanging, but it enabled a deeper insight into the structure - mechanics relation of carbon fibres: By combining mechanical parameters such as the activation volume and activation energy with the structural parameters, the arrangement of the carbon atoms within the crystallites as well as the orientation and the size of the crystallites themselves, it was possible to distinguish between different models for stress graphitisation. The model, which is consistent with the experimental results, is straightening of carbon planes, growth of crystallites, and in particular the rearrangement of cross-linking planes, which are not oriented in the fibre axis.

7.1 Outlook

- Investigation of the boundaries of a graphene sheets in the graphene crystallite or the cross section of a carbon fibre with TEM and EELS, respectively to reveal the positions nitrogen is bonded to. With high resolution TEM it is possible to observe defects in single graphene layers [36] and observe in-situ defect formation. With EELS one can obtain information on the chemical composition of a thin sample, but also on chemical bonding and electronic structure [30]. Recent steps forward in instrumental development allow to record EELS spectra with high energy resolution and spatial resolution in the range of 1 nm.

- An interesting question is the structure at the grain boundaries, in particular, if dangling bonds or cross-linking atoms are present. EELS is suited to determine the bond ratio of sp^2 (graphitic bonds) and sp^3 (diamond-like bonds), because they show a clearly distinguishable signature [20].

7.1. Outlook

- A Raman microprobe can be used to investigate the structural integrity of the single carbon fibres [20]. Raman spectroscopy allows to determine the degree of structural disorder of the fibres by calculating the ratio of the integrated intensity of the D-band ($1360\,\text{cm}^{-1}$) and the G-band ($1580\,\text{cm}^{-1}$) [80, 82]. This describes the amount of disorganized material [47].

- A simulation of the nitrogen loss using information from theoretical works [10, 18] will possibly explain the question of dangling bonds.

Appendix

Construction designs of the holder for the LVDT, the water cooled blocks, one part of a jaw, several adaptors and the heat shield of the fibre bundle test equipment:

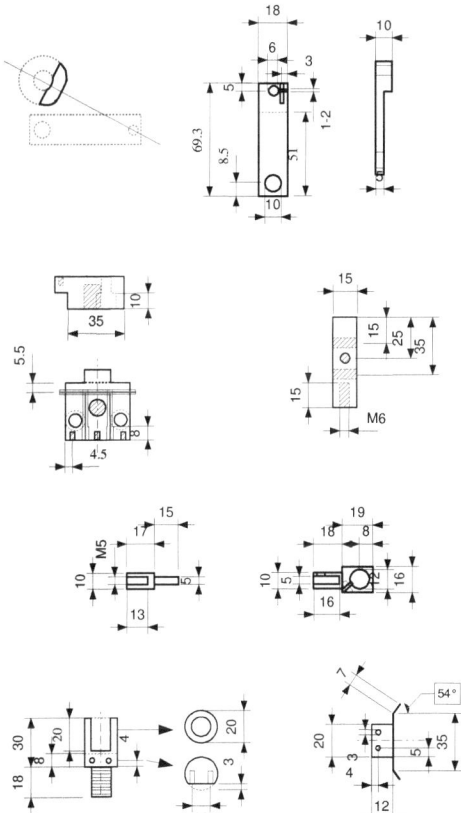

Construction designs of a feed through, a vacuum sealed screw to fix the tension testing device and the top flange of the fibre bundle test equipment:

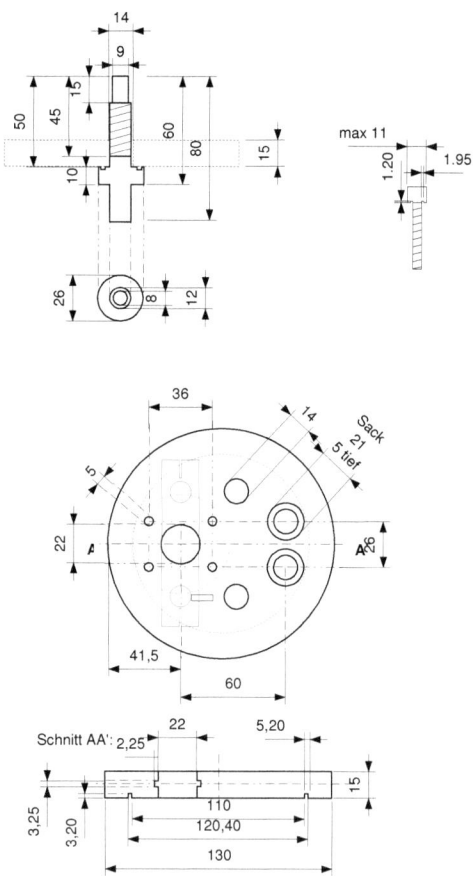

Construction designs of the sample holder of the single fibre test equipment:

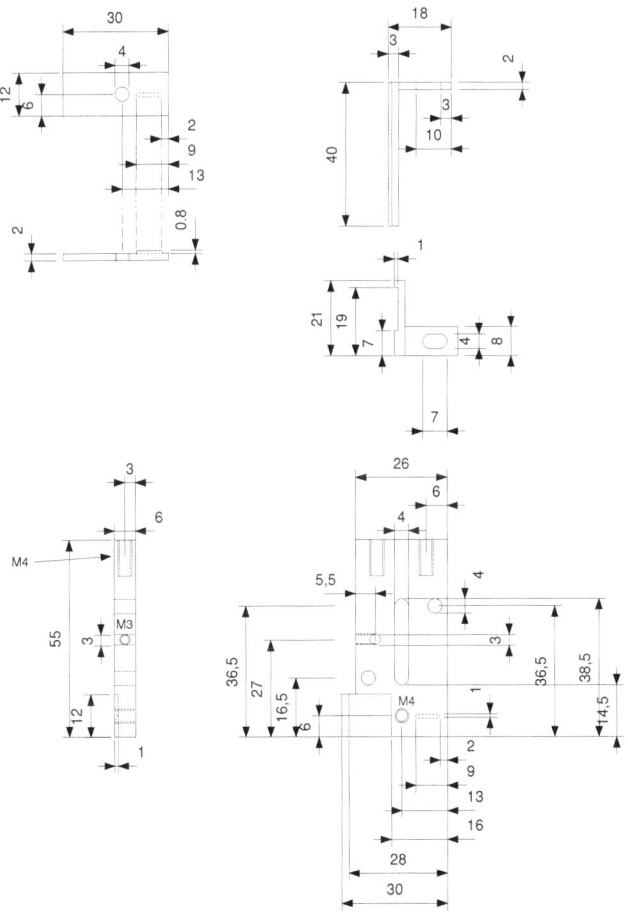

List of Figures

2.1	Two allotropes of carbon	6
2.2	Graphite lattice	7
2.3	Turbostratic graphite	7
2.4	Fibre section	8
2.5	Structure of a carbon fibre	9
2.6	Pores in a carbon fibre	10
2.7	Fibre orientation during the production	11
2.8	Creep curve, scheme	13
2.9	Laue criterion	16
2.10	Transmission $10\mu m$ Al-foil	18
2.11	Diffraction pattern of a carbon fibre	20
2.12	Scheme of the 002 reflections and the 10 band	21
2.13	SAXS slope with Porod region	26
3.1	Scattering image of a carbon fibre bundle	28
3.2	The vessel	31
3.3	Calculate the distance - scheme	32
3.4	Sample holder scheme	33
3.5	Load train	34
3.6	Grips scheme	34
3.7	Grips foto	35
3.8	Temperature distribution along HTA5131	39
3.9	Calibration of the pyrometer, setup	40
3.10	Calibration of the pyrometer	42
3.11	Scattering of the heat shield	44
3.12	Beam-stop	45
3.13	Creep threshold	48
3.14	Resistivity versus fibre number	52
3.15	Measurements 1 and 2	59
3.16	Measurements 4 and 6	60
3.17	Measurements 7 and 8	61

List of Figures

3.18	Measurements 9 and 10	62
3.19	Measurements 11 and 12	63
3.20	Measurements 13 and 14	64
3.21	Measurements 15 and 16	65
3.22	Fit to determine ν	67
3.23	Fit to determine n	69
3.24	Hwhm depending on temperature	70
3.25	Hwhm depending on stress	71
3.26	d_{002} depending on the temperature	72
3.27	d_{002} depending on stress	74
3.28	Radius of gyration depending on the temperature	75
3.29	R_g depending on stress	76
3.30	Crystallite size depending on the temperature	77
3.31	L_c depending on stress	79
3.32	Thermal expansion, data	80
4.1	Scattering image of a single fibre	82
4.2	Single fibre setup	84
4.3	Scheme single fibre setup	85
4.4	Scheme single fibre setup	86
4.5	Resistivity of a single fibre	90
4.6	Single fibre temperature calibration	92
4.7	Calibration of single fibre heating power	94
4.8	Variation of the fibre diameter d	95
4.9	Fibre diameter over normalized resistivity	97
4.10	Error of single fibre temperature	98
4.11	d_{002}, single fibre	105
4.12	Hwhm of the layers, single fibre	106
4.13	HWHM of the pores, single fibre	107
4.14	Scheme crystallographic directions	107
4.15	Thermal expansion d_{002}, FT500 single fibre	108
4.16	Thermal expansion $d_{10,\parallel}$, single fibre	109
4.17	Thermal expansion $d_{10,\perp}$, single fibre	111
5.1	Post creep, $hwhm$	114
5.2	Post creep, R_g	115
6.1	Thermal expansion, discussion	118
6.2	Thermal expansion, in plane, MPP fibres	119
6.3	Thermal expansion at 310 °C	120
6.4	Azimuthal intensity distribution of the 002 peak of M15	121

6.5	Radial intensity distribution of the 002 peak of M9	122
6.6	Structural change, MPP - PAN	123
6.7	Coherent scattering crystallite	124
6.8	Activation volume	125
6.9	Scheme for grain boundary sliding	126
6.10	Scheme for grain boundary sliding	126

List of Tables

2.1	Mechanical properties of carbon fibres	12
2.2	Typical X-ray wavelengths	17
3.1	Power to heat the fibre bundle	36
3.2	Calibration of the pyrometer, 101.9 %	42
3.3	Heat shield foils - properties	44
3.4	Companies found useful	47
3.5	Estimation of the fibre number with different methods	52
3.6	Bundle experiments - overview	57
3.7	Bundle experiments - parameters	58
3.8	Strain rates to determine Q	67
3.9	Strain rates to determine n	68
3.10	HWHM at room temperature	70
3.11	d_{002} at room temperature	73
3.12	R_g at room temperature	76
3.13	L_c at room temperature	78
3.14	Chemical analysis	81
4.1	Calibration of the Sigraflex compound mass	88
4.2	Single fibre calibration data	96
4.3	Single fibre measurement data	102
4.4	Thermal expansion coefficients	112
6.1	Thermal expansion coefficients, discussion	117
6.2	Dependence of the structural parameters	125

Bibliography

[1] Internationale Grundwertreihen für Thermoelemente nach IEC 584-1, Grenzabweichungen nach IEC 584-2, Prosp.-Nr. 8006, Ausgabe 10.90.

[2] *Landoldt Börnstein*, volume 2(1). Springer-Verlag, Berlin, 1971.

[3] Project report. EU-Project VaFTeM (G6RD-CT-2001-00523) work package 3, 2004.

[4] Homepage. http://physics.nist.gov, 2006.

[5] Homepage. http://en.wikipedia.org/wiki/Allotropes_of_carbon, 2008.

[6] Homepage. http://www.periodensystem.info, 2008.

[7] Homepage. http://www.tohotenaxamerica.com/products/pls017.php, 2008.

[8] Homepage. http://www.univie.ac.at/Mikrolabor/chns.htm, 2008.

[9] G. Beaucage. Approximations leading to a unified exponential/power-law approach to small- angle scattering. *Journal of Applied Crystallography*, 28:717–728, 1995.

[10] E.A. Belenkov. Modeling of formation of a crystal structure in a carbon fiber. *Crystallography Reports*, 44(5):749–754, 1999.

[11] S.C. Bennett and D.J. Johnson. Fifth london international carbon and graphite conference, society of chemical industry. 1978. London.

[12] T. Bischoff. A 20-year partnership with the composites industry. *Knitting Technology*, 25(7-8):23, 2003.

[13] O.L. Blakslee, D.G. Proctor, E. Seldin, G.B. Spence, and T. Weng. Elastic constants of compression-annealed pyrolytic graphite. *Journal of Applied Physics*, 41(8):3373–3382, 1970.

Bibliography

[14] G. Boitier, J. Vicens, and J.L. Chermant. Nanostructure study by TEM and HREM of carbon fibres in C_f-SiC composites. *Journal of Materials Science Letters*, 16:1402–1405, 1997.

[15] W.R. Cannon and T.G. Langdon. Review, creep of ceramics, part 1. *Journal of Materials Science*, 18:1–50, 1983.

[16] W.R. Cannon and T.G. Langdon. Review, creep of ceramics, part 2. *Journal of Materials Science*, 23:1–20, 1988.

[17] M. Chenaf. Carbon fibre to reinforce buildings. *JEC Composites Magazine*, 42(15):29, 2005.

[18] B.N. Cox, N. Sridhar, and Argento C.R. A bridging law for creeping fibres. *Acta Matererialia*, 48:4137–4150, 2000.

[19] J. Dienes. Mechanism for self-diffusion in graphite. *Journal of Applied Physics*, 23(11):1194–200, 1953.

[20] M.S. Dresselhaus, G. Dresselhaus, K. Sugihara, I.L. Spain, and Goldberg H.A. *Graphite fibers and filaments*. Springer-Verlag, Berlin Heidelberg, 1988.

[21] T. El Maaddawy, K. Soudki, and T. Topper. Performance evaluation of carbon fiber-reinforced polymer-repaired beams under corrosive environmental conditions. *ACI Structural Journal*, 104(1):3–11, 2007.

[22] F. Entwisle. Thermal expansion of pyrolytic graphite. *Physics Letters*, 2(5):236–238, 1962.

[23] L.A. Feigin and D.I. Svergun. *Structure analysis by small-angle X-ray and neutron scattering*. Plenum Press, New York, 1987.

[24] E. Fitzer. Carbon fibres - the miracle material for temperatures between 5 and 3000 K. *High Temperatures - High Pressures*, 18:479–508, 1986.

[25] E. Fitzer and M. Heym. High-temperature mechanicla properties of carbon and graphite. *High Temperatures - High Pressures*, 10:29–66, 1978.

[26] E. Fitzer and L.M. Manocha. *Carbon reinforcements and carbon/carbon composites*. Springer Verlag, Berlin, 1998.

[27] U. Förstner. *Umweltschutztechnik, 2. Auflage*. Springer, Berlin, 1991.

[28] A.K. Geim and K.S. Novoselov. The rise of graphene. *Nature Materials*, 6(3):183–191, 2007.

[29] O. Glatter and O. Kratky. *Small angle X-ray scattering*. Academic press, London, 1982.

[30] W. Grogger, G. Kothleitner, and F. Hofer. Advantages of a monochromated transmission electron microsocope for solid state physics. *Proceedings of the 56th ÖPG meeting*, page 36, 2006.

[31] A. Guinier. *X-ray crystallographic technology*. Hilger and Watts Ltd., London, 1952.

[32] A. Guinier and G. Fournet. *Small-angle scattering of X-rays*. Wiley, London, New York, 1955.

[33] A. Gupta, I.R. Harrision, and J. Lahijani. Small-angle X-ray scattering in carbon fibres. *Journal of Applied Crystallography*, 27(4):627–636, 1994.

[34] H.S. Gupta, P. Fratzl, M. Kerschnitzki, G. Benecke, W. Wagermaier, and H.O.K. Kirchner. Evidence for an elementary process in bone plasticity with an activation enthalpy of 1eV. *Journal of the Royal Society Interface*, 4(13):277–282, 2007.

[35] J. W. Harrison. Absolute measurements of the coefficient of thermal expansion of elastic constants of compression-annealed pyrolytic graphite from room temperature to 1200 K and a comparison with current theory. *High Temperature - High Pressures*, 9:211–229, 1977.

[36] A. Hashimoto, K. Suenaga, A. Gloter, K. Urita, and S. Iijima. Direct evidence for atomic defects in graphene layers. *Nature*, 430:870–873, 2004.

[37] T. Helmer, H. Peterlik, and Kromp K. Coating of carbon fibers - the strength of the fibers. *Journal of the American Ceramic Society*, 78(1):133–136, 1995.

[38] B.L. Henke, E.M Gullikson, and Davis J.C. X-ray interactions: photoabsorption, scattering, transmission, and reflection at E=50-30000 eV, Z=1-92. *Atomic Data and Nuclear Data Tables*, 54(2):181–342, 1993.

[39] Y. Huang and R.J. Young. Microstructure and mechanical properties of pitch-based carbon fibres. *Journal of Materials Science*, 29(15):4027–4036, 1994.

Bibliography

[40] D.J. Johnson. Structure-property relationships in carbon fibres. *Journal of Physics D: Applied Physics*, 20(3):286–291, 1987.

[41] A. Kanter. Diffusion of carbon atom in natural graphite crystal. *Physical Review*, 107(3):655–663, 1957.

[42] C. Kittel. *Einführung in die Festkörperphysik*. R. Oldenburg Verlag München, Wien, 1969.

[43] K. Kogure, G. Sines, and G. Lavin. Microstructure and texture of pitch-based carbon fibers after creep deformation. *Carbon*, 32(8):1469–1484, 1994.

[44] K. Kogure, G. Sines, and G. Lavin. Structural studies of postcreep, PAN-based, carbon filaments. *Carbon*, 32(4):715–726, 1994.

[45] K. Kogure, G. Sines, and G. Lavin. Creep behavior of a pitch-based carbon filament. *Journal of the American Ceramic Society*, 79(1):46–50, 1996.

[46] J.G. Lavin, K. Kogure, and G. Sines. Mechanical and physical properties of post-creep, pitch-based carbon filaments. *Journal of Materials Science*, 30:2352–2357, 1995.

[47] F. Liu, H. Wang, L. Xue, L. Fan, and Z. Zhu. Effect of microstructure on the mechanical properties of PAN-based carbon fibres during high-temperature graphitization. *Journal of Material Science*, 43:4316–4322, 2008.

[48] D. Loidl. Einfluß einer Hochtemperaturbehandlung auf die mechanischen Eigenschaften von Kohlenstoffasern. Master's thesis, Universität Wien, Vienna, 2000.

[49] D. Loidl, O. Paris, M. Burghammer, C. Riekel, and H. Peterlik. Direct observation of nanocrystallite buckling in carbon fibers under bending load. *Physical Review Letters*, 95:225501, 2005.

[50] D. Loidl, O. Paris, H. Rennhofer, M. Müller, and H. Peterlik. Skin-core structure and bimodal Weibull distribution of the strength of carbon fibres. *Carbon*, 45:2801–2805, 2007.

[51] D. Loidl, H. Peterlik, and K. Kromp. Statistical parameters of fibres for endless-fibre reinforced caremic matrix composites evaluated by the fibre-bundle tension test. *Proceedings of The 5th International*

Symposion on Brittle Matrix Composites, pages 524–536, 1997. Bigraf, Warszawa, Poland.

[52] D. Loidl, H. Peterlik, M. Müller, Riekel C., and O. Paris. Elastic moduli of nanocrystallites in carbon fibers measured by in-situ X-ray microbeam diffraction. *Carbon*, 41:563–570, 2003.

[53] J.C. Meyer, A.K. Geim, M.I. Katsnelson, K.S. Novoselov, T.J. Booth, and S. Roth. The structure of suspended graphene sheets. *Nature*, 446(7131):60–63, 2007.

[54] E.W. Nuffield. *X-ray diffraction methods*. John Wiley & Sons, Inc., New York, 1966.

[55] A. Oberlin. Carbonization and graphitization. *Carbon*, 22(6):521–541, 1984.

[56] A.A. Ogale, C. Lin, D.P. Anderson, and K.M. Kearns. Orientation and dimensional changes in mesophase pitch-based carbon fibers. *Carbon*, 40:1309–1319, 2002.

[57] O. Paris, D. Loidl, and H. Peterlik. Texture of PAN- and pitch-based carbon fibers. *Carbon*, 40:551–555, 2002.

[58] O. Paris, D. Loidl, H. Peterlik, M. Müller, H. Lichtenegger, and P. Fratzl. The internal structure of single carbon fibers determined by simultaneous small- and wide-angle scattering. *Journal of Applied Crystallography*, 33:695–699, 2000.

[59] O. Paris and H. Peterlik. Carbon fibres. Book article, in print, 2008.

[60] R. Perret and W. Ruland. Single and multiple X-ray small-angle scattering of carbon fibres. *Journal of Applied Crystallography*, 2:209–218, 1969.

[61] R. Perret and W. Ruland. The microstructure of PAN-base carbon fibres. *Journal of Applied Crystallography*, 3:525–532, 1970.

[62] H. Peterlik, P. Fratzl, and K. Kromp. Pore structure of carbon/carbon composites studied by small-angle x-ray scattering. *Carbon*, 32(5):939–945, 1994.

[63] H. Peterlik and D. Loidl. Bimodal strenght distriubtions and flaw populations of ceramics and fibres. *Engineering Fracture Mechanics*, 68:253–261, 2001.

Bibliography

[64] C. Reder. Einfach-, Mehrfach-, Zugfasertests an Carbonfasern zur Verstärkung von keramischen Verbundwerkstoffen mit keramischer Matrix. Master's thesis, Universität Wien, Vienna, 2002.

[65] C. Reder, D. Loidl, S. Puchegger, D. Gitschthaler, H. Peterlik, K. Kromp, G. Khatibi, A. Betzwar-Kotas, P. Zimprich, and B. Weiss. Non-contacting strain measurements of ceramic and carbon single fibres by using the laser-speckle method. *Composites: Part A*, 34:1029–1033, 2003.

[66] H. Rennhofer. Kriechfestigkeit keramischer Fasern bei hohen Temperaturen, Mechankik und Struktur. Master's thesis, Universität Wien, Vienna, 2003.

[67] H. Rennhofer, D. Loidl, J. Brandstetter, K. Kromp, R. Weiss, and H. Peterlik. Structural change of carbon-fibres at high temperatures under load. *Fatigue & Fracture of Engineering Materials & Structures*, 29(2):167–172, 2006.

[68] T. Roberts. *The carbon fibre industry*. Materials Technology Publications, UK, 2006.

[69] F. Rozploch and W. Marciniak. Radial thermal expansion of carbon fibres. *High Temperatures - High Pressures*, 18(5):585–587, 1986.

[70] W. Ruland. Apparent fractal dimensions obtained from small-angle scattering of carbon materials. *Carbon*, 39(2):323–324, 2001.

[71] C. Sauder, J. Lamon, and R. Pailler. Thermomechanical properties of carbon fibres at high temperatures (up to 2000 °C). *Composites Science and Technology*, 62:499–504, 2002.

[72] C. Sauder, J. Lamon, and R. Pailler. The tensile behavior of carbon fibres at high temperatures (up to 2400 °C). *Carbon*, 42:715–725, 2004.

[73] G. Savage. *Carbon-carbon composites*. Chapman&Hall, New York, 1993.

[74] P.W. Schmidt. Small-angle scattering studies of disordered, porous and fractal systems. *Journal of Applied Crystallography*, 24:414–435, 1991.

[75] G. Sines, Z. Yang, and B.D. Vickers. Effect of the matrix and matrix bonding on the creep behavior of a unidirectional carbon-carbon composite. *Journal of the American Ceramic Society*, 72(1):54–59, 1989.

[76] G. Sines, Z. Yang, and B.D. Vickers. Friction stress for carbon composites and carbon yarn during high temperature creep. *Acta Metallurgica*, 37(10):2673–2680, 1989.

[77] T. Tagawa and T. Miyata. Size effect on tensile strength of carbon fibers. *Materials Science and Engineering A*, 238:336–342, 1997.

[78] A. Takaku and M. Shioya. X-ray measurements and the structure of polyacrylonitrile- and pitch-based carbon fibres. *Journal of Materials Science*, 25:4873–4879, 1990.

[79] C.A. Taylor, M.F. Wayne, and W.K.S. Chiu. Heat treatment of thin carbon films and the effect of residual stress, modulus, thermal expansion and microstructure. *Carbon*, 41:1867–1875, 2003.

[80] F. Tuinstra and J.L. Koenig. Raman spectrum of graphite. *Journal Of Chemical Physics*, 53(3):1126–1130, 1970.

[81] P. J. Walsh. Carbon fibers. *ASM Handbook, Composites*, 21:35–40, 2001.

[82] G.A. Zickler, B. Smarsly, N. Gierlinger, H. Peterlik, and O. Paris. A reconsideration of the relationship between the crystallite size L_a of carbons determined by X-ray diffraction and Raman spectroscopy. *Carbon*, 44:3239–3246, 2006.

Acknowledgement

Thanks to all people who directly or indirectly supported my work.

I had the great opportunity to work with: Herwig Peterlik (who was also the supervisor of this thesis), Dieter Loidl, Karl Kromp, Roland, Hans, Stephan, Oskar, Daniel, Silvia, Manuel.
I was also supported by: R. Weiss (Schunk GmbH), M. Kutschera (VIAS), J. Theiner (Mikroanalytisches Laboratorium), S. Siegel (BESSY).
There was a lot of private support and help: Kathy, MxMcK, Marcello-san, Oliver, Michi (voc), Sara, Martin, Ilse, Gottfried, Monika, Philipp.
Furthermore the support of the Austrian Science Fund FWF and the support of the University of Vienna is gratefully acknowledged.

VDM Verlagsservicegesellschaft mbH

Die VDM Verlagsservicegesellschaft sucht für wissenschaftliche Verlage abgeschlossene und herausragende

Dissertationen, Habilitationen, Diplomarbeiten, Master Theses, Magisterarbeiten usw.

für die kostenlose Publikation als Fachbuch.

Sie verfügen über eine Arbeit, die hohen inhaltlichen und formalen Ansprüchen genügt, und haben Interesse an einer honorarvergüteten Publikation?

Dann senden Sie bitte erste Informationen über sich und Ihre Arbeit per Email an *info@vdm-vsg.de*.

Sie erhalten kurzfristig unser Feedback!

VDM Verlagsservicegesellschaft mbH
Dudweiler Landstr. 99
D - 66123 Saarbrücken

Telefon +49 681 3720 174
Fax +49 681 3720 1749

www.vdm-vsg.de

Die VDM Verlagsservicegesellschaft mbH vertritt

Printed by Books on Demand GmbH, Norderstedt / Germany